STUDENT WORKBOOK

VOLUME 2

COLLEGE PHYSICS

A STRATEGIC APPROACH

KNIGHT • JONES • FIELD

RANDALL D. KNIGHT
CALIFORNIA POLYTECHNIC STATE UNIVERSITY,
SAN LUIS OBISPO

JAMES H. ANDREWS
YOUNGSTOWN STATE UNIVERSITY

San Francisco Boston New York
Capetown Hong Kong London Madrid Mexico City
Montreal Munich Paris Singapore Sydney Tokyo Toronto

Editorial Director: Adam Black
Sponsoring Editor: Alice Houston
Development Manager: Michael Gillespie
Project Editor: Martha Steele
Managing Editor: Corinne Benson
Production Supervisor: Shannon Tozier
Production Service and Compositor: WestWords, Inc.
Illustrations: Precision Graphics
Text Design: Seventeenth Street Studios and WestWords, Inc.
Cover Design: Yvo Riezebos and Seventeenth Street Studios
Manufacturing Manager: Pam Augspurger
Text and Cover Printer: Bind-Rite Graphics
Cover Image: Ken Wilson; Papilio/CORBIS

ISBN: 0-8053-8209-7 Volume 1
ISBN: 0-8053-0626-9 Volume 2

1 2 3 4 5 6 7 8 9 10—BRG—10 09 08 07 06
www.aw-bc.com

Table of Contents

Part V Optics

Chapter 17 Wave Optics 17-1
Chapter 18 Ray Optics 18-1
Chapter 19 Optical Instruments 19-1

Part VI Electricity and Magnetism

Chapter 20 Electric Forces and Fields 20-1
Chapter 21 The Electric Potential 21-1
Chapter 22 Current and Resistance 22-1
Chapter 23 Circuits 23-1
Chapter 24 Magnetic Fields and Forces 24-1
Chapter 25 Electromagnetic Induction and Electromagnetic Waves 25-1
Chapter 26 AC Circuits 26-1

Part VII Modern Physics

Chapter 27 Relativity 27-1
Chapter 28 Quantum Physics 28-1
Chapter 29 Atoms and Molecules 29-1
Chapter 30 Nuclear Physics 30-1

Notes

Preface

It is highly unlikely that one could learn to play the piano by only reading about it. Similarly, reading physics from a textbook is not the same as doing physics. To develop your ability to do physics, your instructor will assign problems to be solved both for homework and on tests. Unfortunately, it is our experience that jumping right into problem solving after reading and hearing about physics often leads to poor "playing" techniques and an inability to solve problems for which the student has not already been shown the solution (which isn't really "solving" a problem, is it?). Because improving your ability to solve physics problems is one of the major goals of your course, time spent developing techniques that will help you do this is well spent.

Learning physics, as in learning any skill, requires regular practice of the basic techniques. That is what this *Student Workbook* is all about. The workbook consists of exercises that give you an opportunity to practice techniques and strengthen your understanding of concepts presented in the textbook and in class. These exercises are intended to be done on a daily basis, right after the topics have been discussed in class and are still fresh in your mind. Successful completion of the workbook exercises will prepare you to tackle the more quantitative end-of-chapter homework problems in the textbook.

You will find that many of the exercises are *qualitative* rather than *quantitative.* They ask you to draw pictures, interpret graphs, use ratios, write short explanations, or provide other answers that do not involve calculations. A few math-skills exercises will ask you to explore the mathematical relationships and symbols used to quantify physics concepts, but do not require a calculator. The purpose of all of these exercises is to help you develop the basic thinking tools you'll later need for quantitative problem solving. It is highly recommended that you do these exercises *before* starting the end-of-chapter problems

One example from Chapter 4 illustrates the purpose of this *Student Workbook.* In that chapter, you will read about a technique called a "free-body diagram" that is helpful for solving problems involving forces. Sometimes students mistakenly think that the diagrams are used by the instructor only for teaching purposes and may be abandoned once Newton's laws are fully understood. On the contrary, professional physicists with decades of problem-solving experience still routinely use these diagrams to clarify the problem and set up the solution. Many of the other techniques practiced in the workbook, such as ray diagrams, graphing relationships, sketching field lines and equipotentials, etc., fall in the same category. They are used at all levels of physics, not just as a beginning exercise. And many of these techniques, such as analyzing graphs and exploring multiple representations of a situation, have important uses outside of physics. Time spent practicing these techniques will serve you well in other endeavors.

You will find that the exercises in this workbook are keyed to specific sections of the textbook in order to let you practice the new ideas introduced in that section. You should keep the text beside you as you work and refer to it often. You will usually find Tactics Boxes, figures, or examples in the textbook that are directly relevant to the exercises. When asked to draw figures or diagrams, you should attempt to draw them so that they look much like the figures and diagrams in the textbook.

Because the exercises go with specific sections in the text, you should answer them on the basis of information presented in *just* that section (and prior sections). You may have learned new ideas in Section 7 of a chapter, but you should not use those ideas when answering questions from Section 4. There will be ample opportunity in the Section 7 exercises to use that information there.

You will need a few "tools" to complete the exercises. Many of the exercises will ask you to *color code* your answers by drawing some items in black, others in red, and perhaps yet others in blue. You need to purchase a few colored pencils to do this. The authors highly recommend that you work in pencil, rather than ink, so that you can easily erase. Few are the individuals who make so few mistakes as to be able to work in ink! In addition, you'll find that a small, easily carried six-inch ruler will come in handy for drawings and graphs.

As you work your way through the textbook and this workbook, you will find that physics is a way of *thinking* about how the world works and why things happen as they do. We will primarily be interested in finding relationships, seeking explanations, and developing techniques to make use of these relationships, only secondarily in computing numerical answers. In many ways, the thinking tools developed in this workbook are what the course is all about. If you take the time to do these exercises regularly and to review the answers, in whatever form your instructor provides them, you will be well on your way to success in physics.

To the instructor: The exercises in this workbook can be used in many ways. You can have students work on some of the exercises in class as part of an active-learning strategy. Or you can do the same in recitation sections or laboratories. This approach allows you to discuss the answers immediately, to answer student questions, and to improvise follow-up exercises when needed. Having the students work in small groups (2 to 4 students) is highly recommended.

Alternatively, the exercises can be assigned as homework. The pages are perforated for easy tear-out, and the page breaks are in logical places so that you can assign the sections of a chapter that you would likely cover in one day of class. Exercises should be assigned immediately after presenting the relevant information in class and should be due at the beginning of the next class. Collecting them at the beginning of class, then going over two or three that are likely to cause difficulty, is an effective means of quickly reviewing major concepts from the previous class and launching a new discussion.

If used as homework, it is *essential* for students to receive *prompt* feedback. Ideally this would occur by having the exercises graded, with written comments, and returned at the next class meeting. Posting fairly detailed answers on a course website also works. Lack of prompt feedback can negate much of the value of these exercises. Placing similar qualitative/graphical questions on quizzes and exams, and telling students at the beginning of the term that you will do so, encourages students to take the exercises seriously and to check the answers.

One of the authors has been successful with assigning *all* exercises in the workbook as homework, collecting and grading them every day through Chapter 4, then collecting and grading them on about one-third of subsequent days on a random basis. The other author uses the exercises in class as immediate practice of the techniques demonstrated in the text and on the chalkboard. Student feedback from end-of-term questionnaires reveals three prevalent attitudes toward the workbook exercises:

i. They think it is an unreasonable amount of work.
ii. They agree that the assignments force them to keep up and not get behind.
iii. They recognize, by the end of the term, that the workbook is a valuable learning tool.

However you choose to use these exercises, they will significantly strengthen your students' conceptual understanding of physics.

Answers to all workbook exercises are provided as pdf files on the *Media Manager* CD-ROM.

Acknowledgments: The authors would like to thank Rebecca L. Sobinovsky and Jared Sterzer.

17 Wave Optics

17.1 What is Light?

1. A light wave travels from vacuum, through a transparent material, and back to vacuum. What is the index of refraction of this material? Explain.

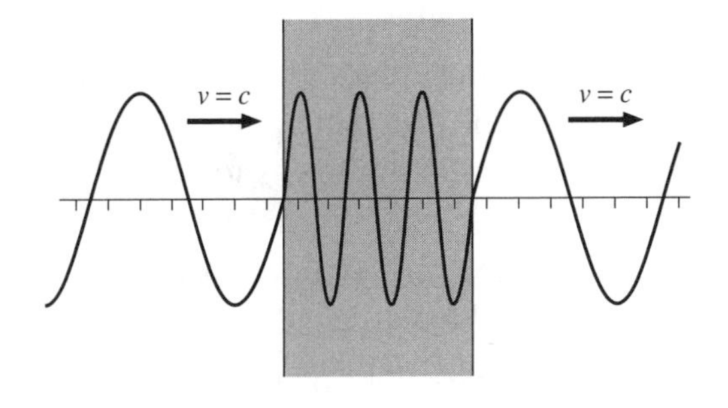

2. A light wave travels from vacuum, through a transparent material whose index of refraction is $n = 2.0$, and back to vacuum. Finish drawing the snapshot graph of the light wave at this instant.

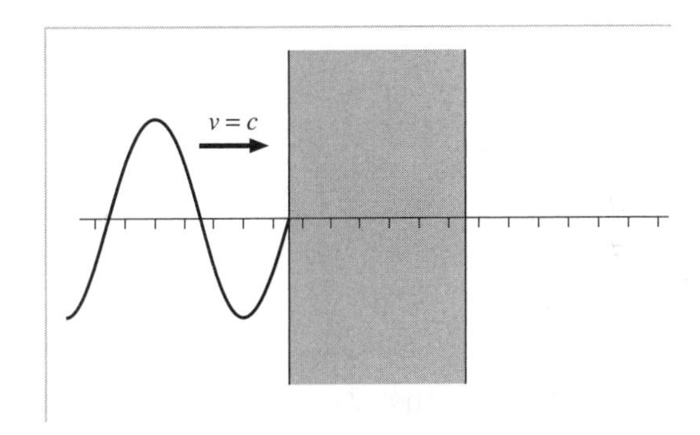

17.2 The Interference of Light

3. The figure shows the light intensity recorded by a piece of film in an interference experiment. Notice that the light intensity comes "full on" at the edges of each maximum, so this is *not* the intensity that would be recorded in Young's double-slit experiment.
   a. Draw a graph of light intensity versus position on the film. Your graph should have the same horizontal scale as the "photograph" above it.

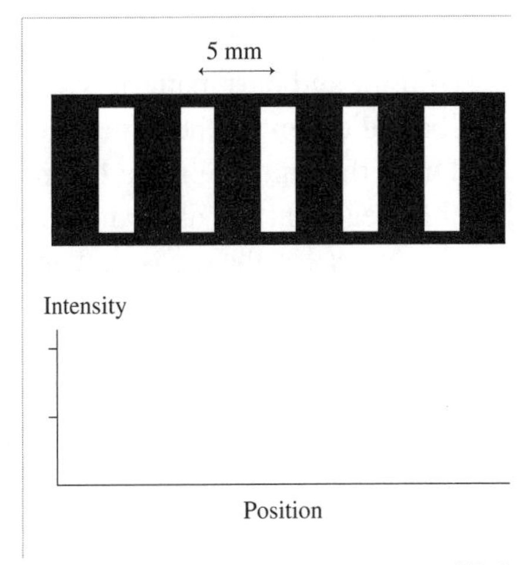

b. Is it possible to tell, from the information given, what the wavelength of the light is? If so, what is it? If not, why not?

4. The graph shows the light intensity on the viewing screen during a double-slit interference experiment.

   a. Draw the "photograph" that would be recorded if a piece of film were placed at the position of the screen. Your "photograph" should have the same horizontal scale as the graph above it. Be as accurate as you can. Let the white of the paper be the brightest intensity and a very heavy pencil shading be the darkest.

   b. Three positions on the screen are marked as A, B, and C. Draw history graphs showing the displacement of the light wave at each of these three positions as a function of time. Show three cycles, and use the same vertical scale on all three.

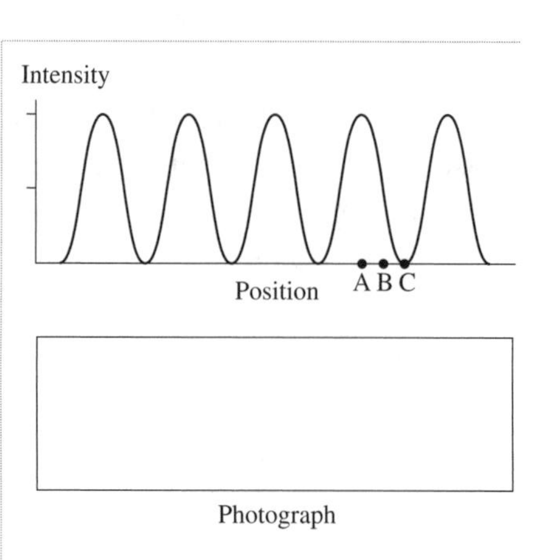

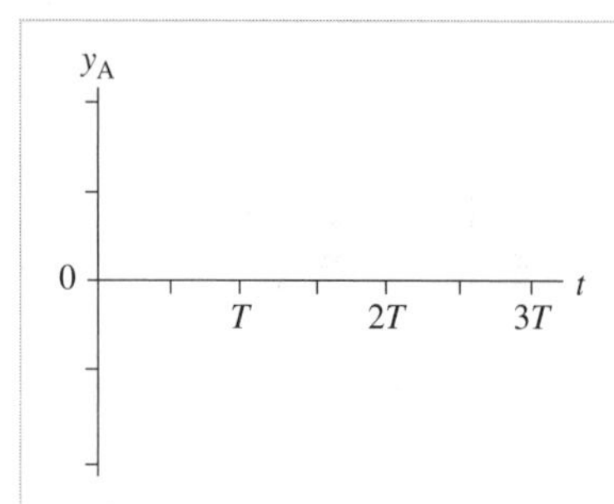

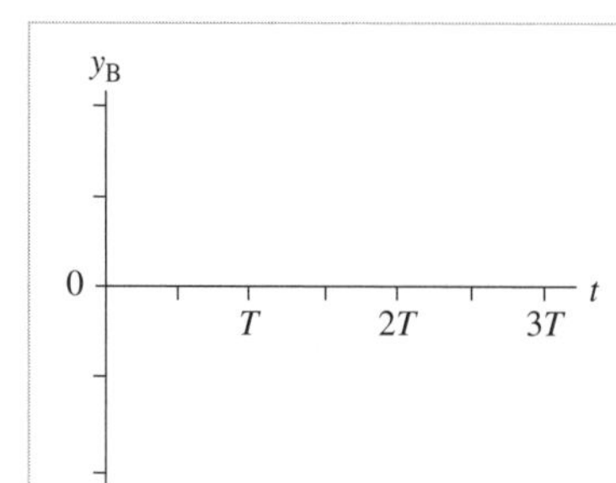

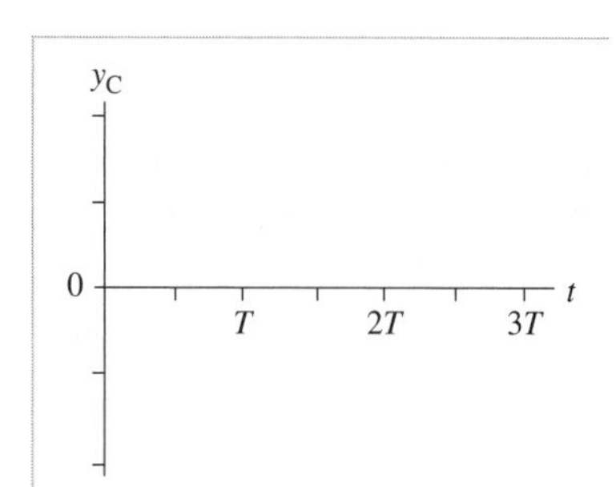

5. In a double-slit experiment, we usually see the light intensity only on a viewing screen. However, we can use smoke or dust to make the light visible as it propagates between the slits and the screen. Consider a double-slit experiment in a smoke-filled room. What kind of light and dark pattern would you see if you looked down on the experiment from above? Draw the pattern on the figure below. Shade the areas that are dark and leave the white of the paper for the areas that are bright.

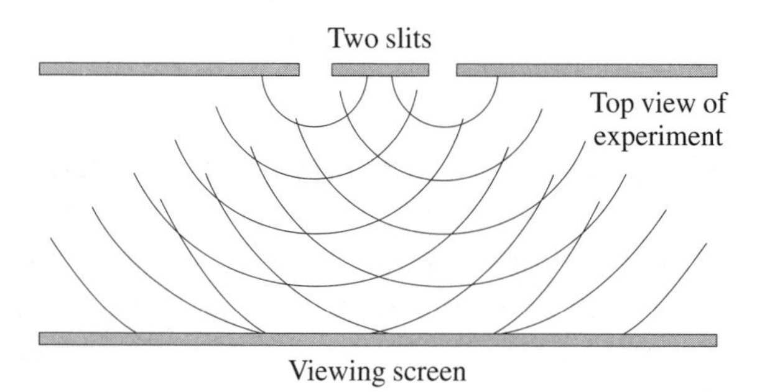

17.3 The Diffraction Grating

6. The figure shows four slits in a diffraction grating. A set of circular wave crests is shown spreading out from each slit. Four wave paths, numbered 1 to 4, are shown leaving the slits at angle θ_1. The dotted lines are drawn perpendicular to the paths of the waves.

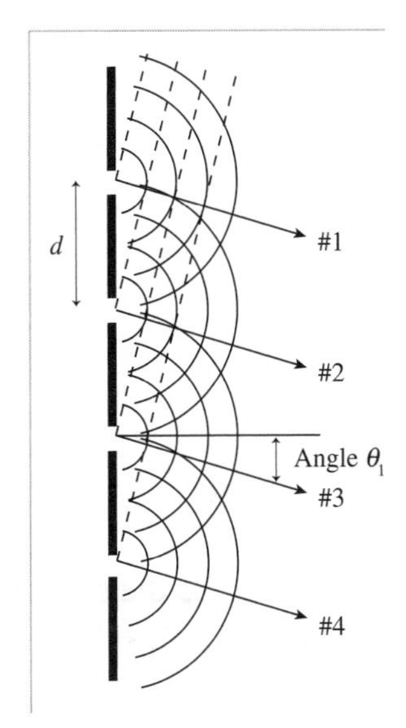

   a. Use a colored pencil or heavy shading to show *on the figure* the extra distance traveled by wave 1 that is not traveled by wave 2.

   b. How many extra wavelengths does wave 1 travel compared to wave 2? Explain how you can tell from the figure.

   c. How many extra wavelengths does wave 2 travel compared to wave 3?

   d. As these four waves combine at some large distance from the grating, will they interfere constructively, destructively, or in between? Explain.

7. Suppose the wavelength of the light in Exercise 6 is doubled. (Imagine erasing every other wave front in the picture.) Would the interference at angle θ_1 then be constructive, destructive, or in between? Explain. Your explanation should be based on the figure, not on some equation.

8. Suppose the slit spacing d in Exercise 6 is doubled while the wavelength is unchanged. Would the interference at angle θ_1 then be constructive, destructive, or in between? Again, base your explanation on the figure.

9. These are the same slits as in Exercise 6. Waves with the same wavelength are spreading out on the right side.
   a. Draw four paths, starting at the slits, at an angle θ_2 such that the wave along each path travels *two* wavelengths farther than the next wave. Also draw dashed lines at right angles to the travel direction. Your picture should look much like the figure of Exercise 6, but with the waves traveling at a different angle. Use a ruler!
   b. Do the same for four paths at angle $\theta_{1/2}$ such that each wave travels *one-half* wavelength farther than the next wave.

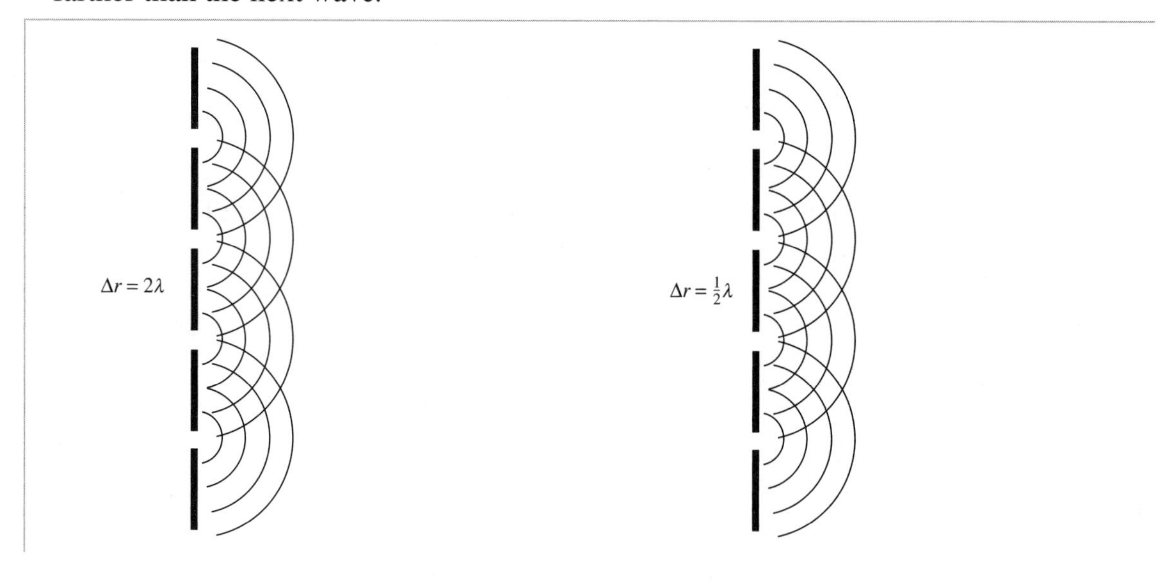

17.4 Thin-Film Interference

10. The figure shows a wave transmitted from air through a thin oil film on water. The film has a thickness of $t = \lambda_{oil}/2$.

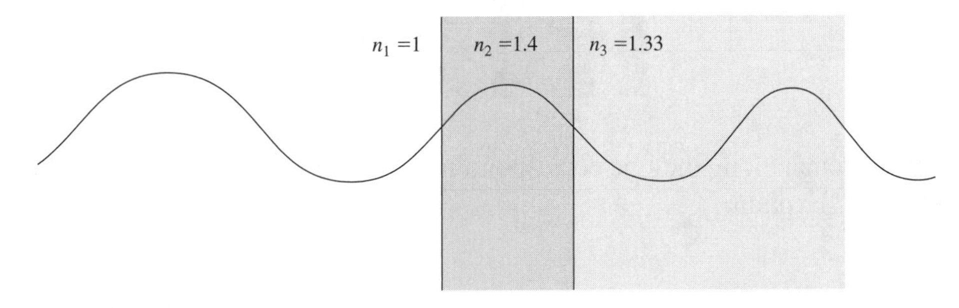

a. Using the indices of refraction that are given, indicate at each interface on the diagram whether the reflected wave (not shown) undergoes a phase change at the boundary.

b. Draw in the reflected wave from the first interface directly below the incident and transmitted waves. Draw the reflected wave from the second interface directly below the reflected wave for the first interface.

c. Do the reflected waves interfere constructively, destructively, or in between? Explain.

11. The figure shows a wave transmitted from air through a thin oil film on water. The film has a thickness of $t = \lambda_{oil}/2$.

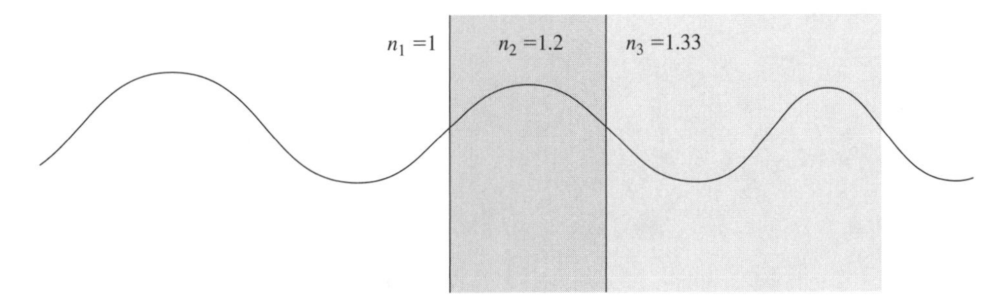

a. Using the indices of refraction that are given, indicate at each interface on the diagram whether the reflected wave (not shown) undergoes a phase change at the boundary.

b. Draw in the reflected wave from the first interface directly below the incident and transmitted waves. Draw the reflected wave from the second interface directly below the reflected wave for the first interface.

c. Do the reflected waves interfere constructively, destructively, or in between? Explain.

12. The figure shows the fringes seen due to a wedge of air between two flat glass plates that touch at one end and are illuminated by light of wavelength $\lambda = 500$ nm.

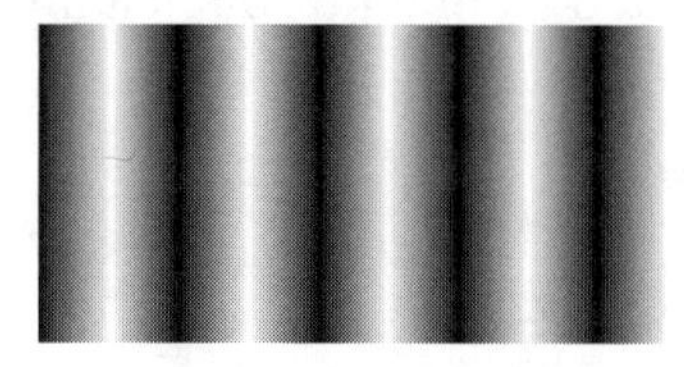

a. Do the plates touch along the line corresponding to the left end or the right end of the fringe pattern shown? Explain.

b. What difference in separation between the glass plates corresponds to the separation between adjacent dark fringes in the fringe pattern shown? Explain.

c. What is the furthest separation between the plates for the pattern shown? Explain.

17.5 Single-Slit Diffraction

13. Plane waves of light are incident on two narrow, closely-spaced slits. The graph shows the light intensity seen on a screen behind the slits.
    a. Draw a graph on the axes below right to show the light intensity on the screen if the right slit is blocked, allowing light to go only through the left slit.
    b. Explain why the graph will look this way.

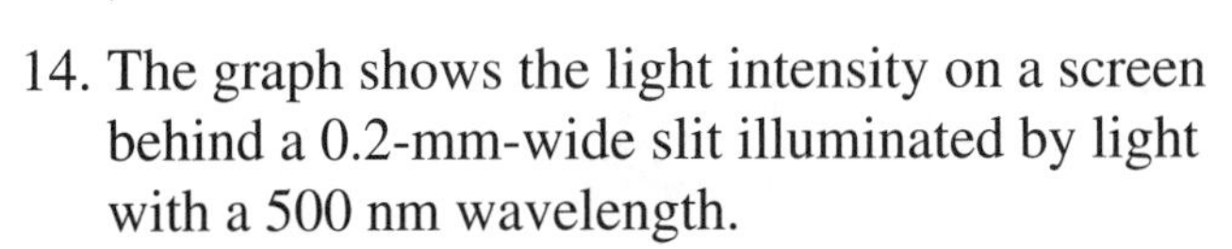

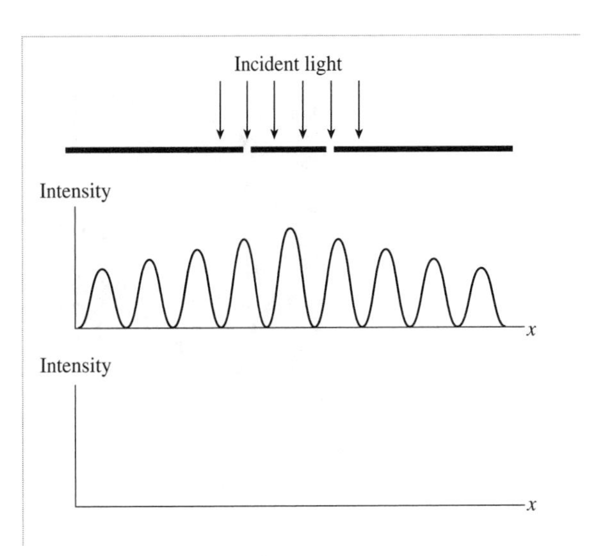

14. The graph shows the light intensity on a screen behind a 0.2-mm-wide slit illuminated by light with a 500 nm wavelength.
    a. Draw a *picture* in the box of how a photograph taken at this location would look. Use the same horizontal scale, so that your picture aligns with the graph above. Let the white of the paper represent the brightest intensity and the darkest you can draw with a pencil or pen be the least intensity.
    b. Using the same horizontal scale as in part a, draw graphs showing the light intensity if
       i. $\lambda = 250$ nm, $a = 0.2$ mm.
       ii. $\lambda = 1000$ nm, $a = 0.2$ mm.
       iii. $\lambda = 500$ nm, $a = 0.1$ mm.
       iv. $\lambda = 500$ nm, $a = 0.4$ mm.

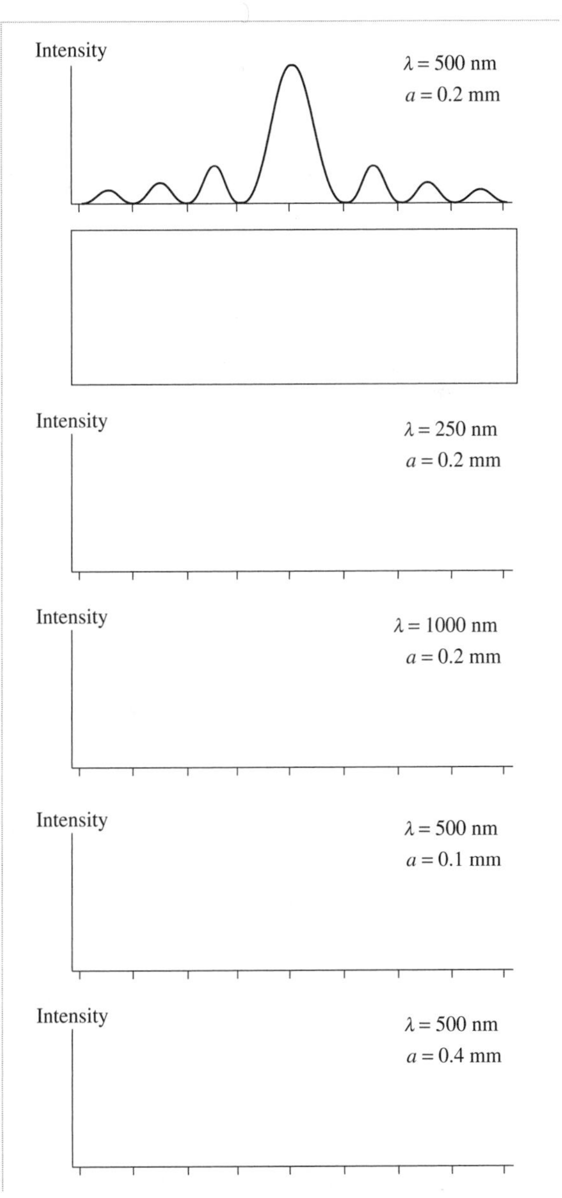

17.6 Circular-Aperture Diffraction

15. This is the light intensity on a viewing screen behind a circular aperture.

a. In the middle box, sketch how the pattern would appear if the wavelength of the light were doubled. Explain.

b. In the far right box, sketch how the pattern would appear if the diameter of the aperture were doubled. Explain.

18 Ray Optics

Note: Please use a ruler or straight edge for drawing light rays.

18.1 The Ray Model of Light

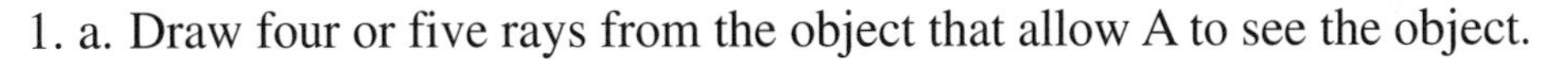

1. a. Draw four or five rays from the object that allow A to see the object.

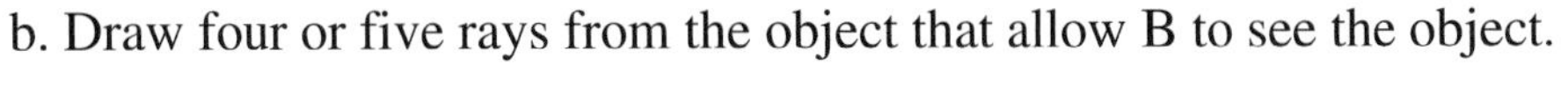

   b. Draw four or five rays from the object that allow B to see the object.

   c. Describe the situations seen by A and B if a piece of cardboard is lowered at point C.

2. a. Draw four or five rays from object 1 that allow A to see object 1.
   b. Draw four or five rays from object 2 that allow B to see object 2.
   c. What happens to the light where the rays cross in the center of the picture?

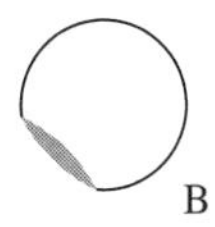

3. A point source of light illuminates a slit in an opaque barrier.
   a. On the screen, sketch the pattern of light that you expect to see. Let the white of the paper represent light areas; shade dark areas. Mark any relevant dimensions.
   b. What will happen to the pattern of light on the screen if the slit width is reduced to 0.5 cm?

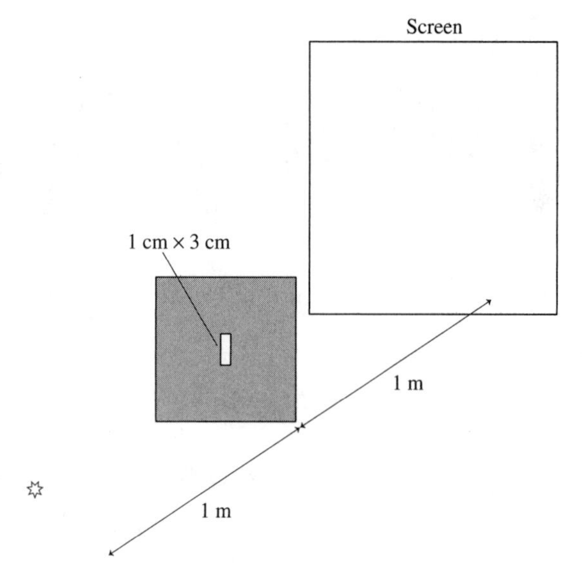

4. In each situation below, light passes through a 1-cm-diameter hole and is viewed on a screen. For each, sketch the pattern of light that you expect to see on the screen. Let the white of the paper represent light areas; shade dark areas.

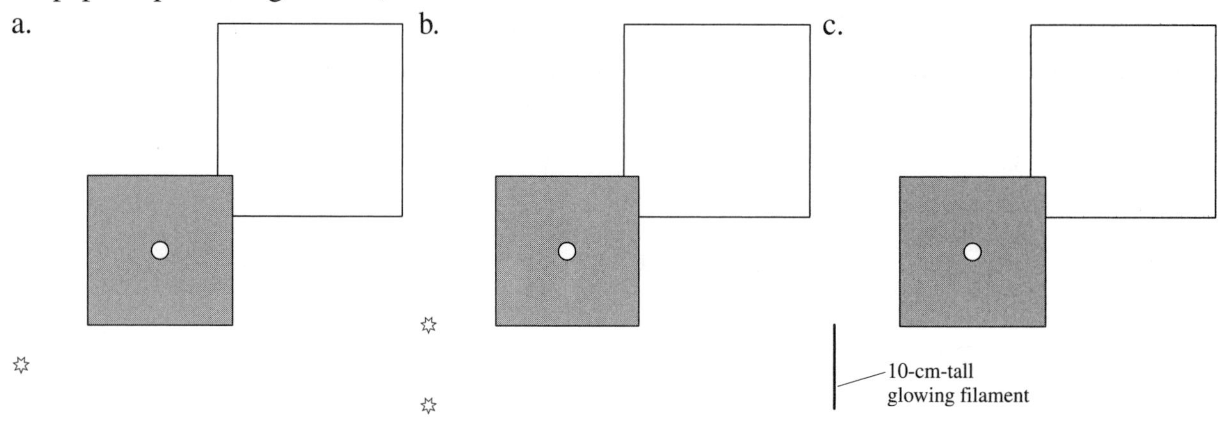

5. Light from an L-shaped bulb passes through a pinhole. On the screen, sketch the pattern of light that you expect to see.

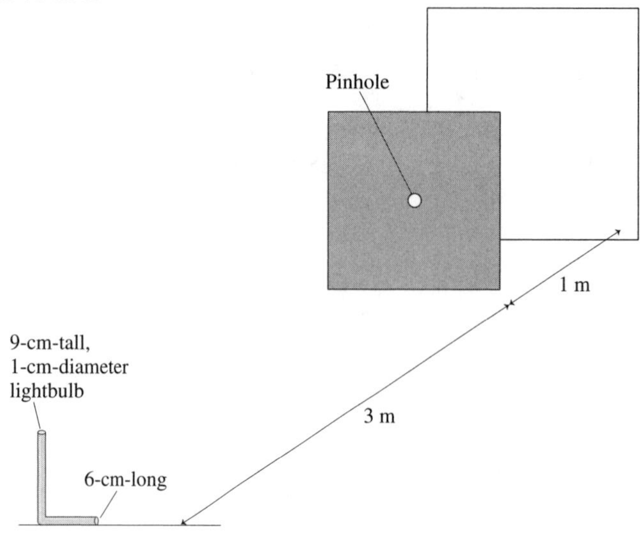

18.2 Reflection

6. a. Draw five rays from the object that pass through points A to E after reflecting from the mirror. Make use of the grid to do this accurately.
   b. Extend the reflected rays behind the mirror.
   c. Show and label the image point.

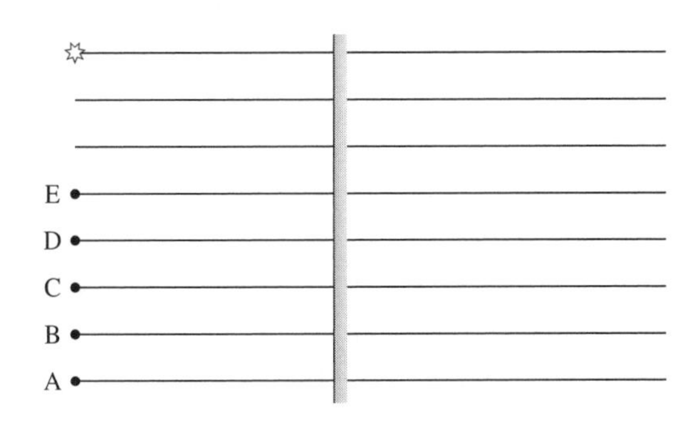

7. a. Draw *one* ray from the object that enters the eye after reflecting from the mirror.
   b. Is one ray sufficient to tell your eye/brain where the image is located?

   c. Use a different color pen or pencil to draw two more rays that enter the eye after reflecting. Then use the three rays to locate (and label) the image point.
   d. Do any of the rays that enter the eye actually pass through the image point?

8. The two mirrors are perpendicular to each other.

   a. Draw a ray directly from the object to point A. Then draw two rays that strike the mirror *very close* to A, one on either side. Use the reflections of these three rays to locate an image point.

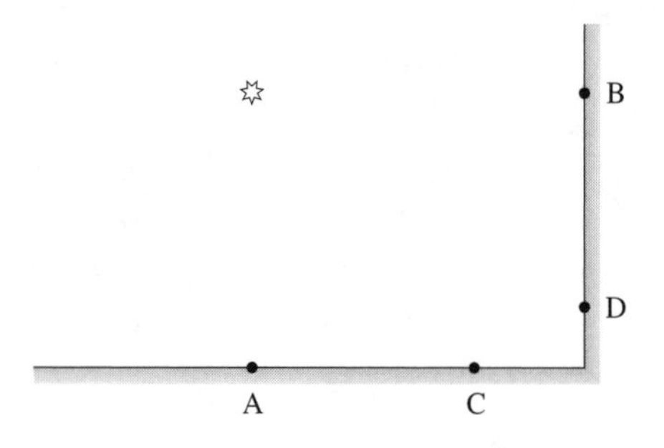

   b. Do the same for points B, C, and D.

   c. How many images are there, and where are they located?

18.3 Refraction

9. Draw seven rays from the object that refract after passing through the seven dots on the boundary.

a.

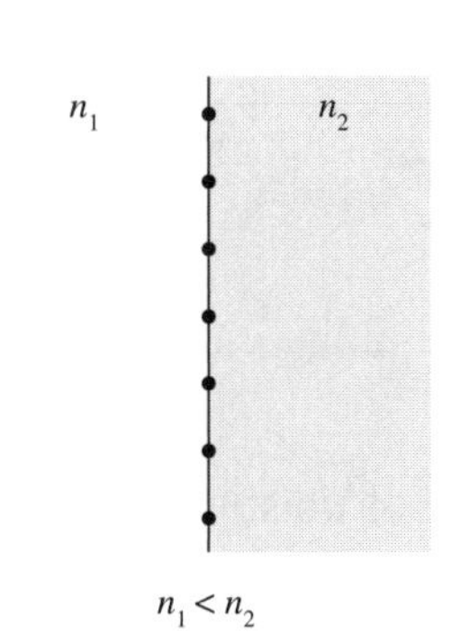

b.

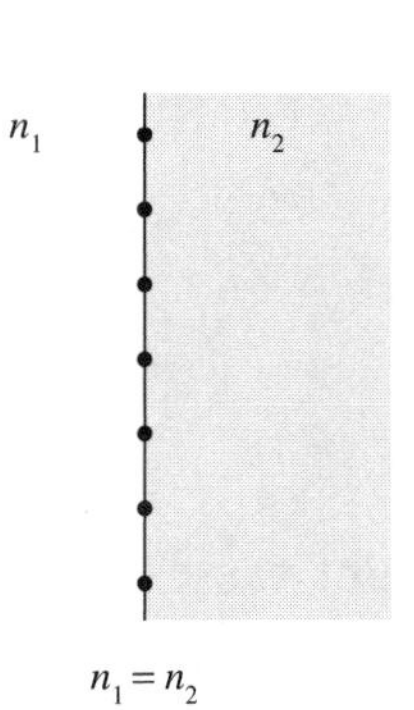

c.

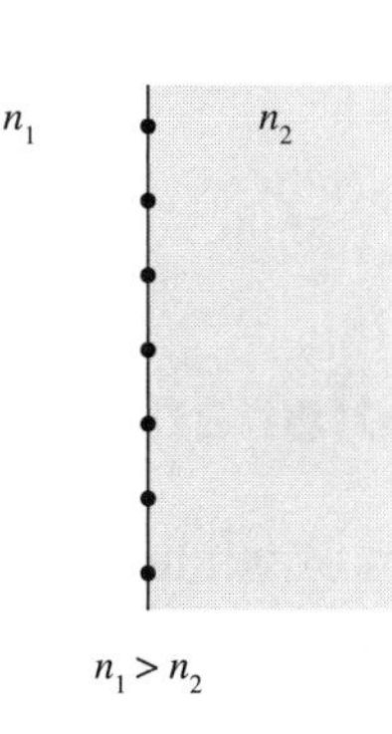

10. Complete the trajectories of these three rays through material 2 and back into material 1. Assume $n_2 < n_1$.

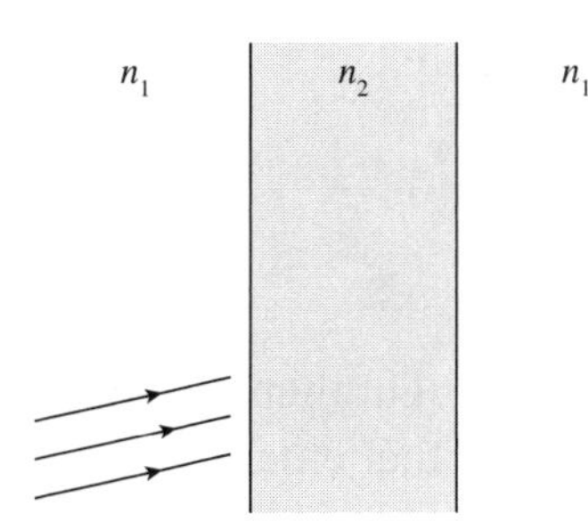

11. The figure shows six conceivable trajectories of light rays leaving an object. Which, if any, of these trajectories are impossible? For each that is possible, what are the requirements of the index of refraction n_2?

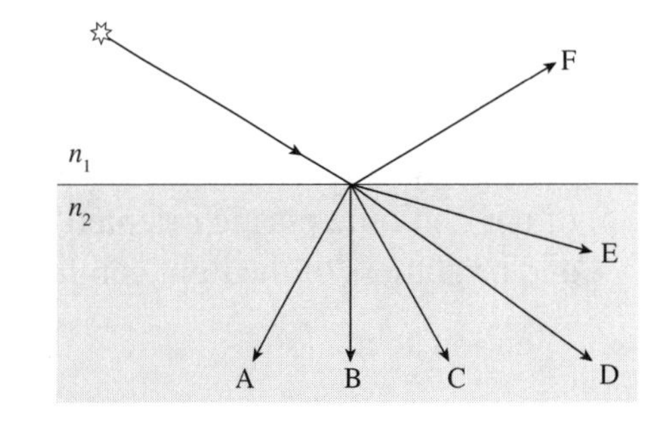

Impossible __________

Requires $n_2 > n_1$ __________

Requires $n_2 = n_1$ __________

Requires $n_2 < n_1$ __________

Possible for any n_2 __________

12. Complete the ray trajectories through the two prisms shown below.

a.

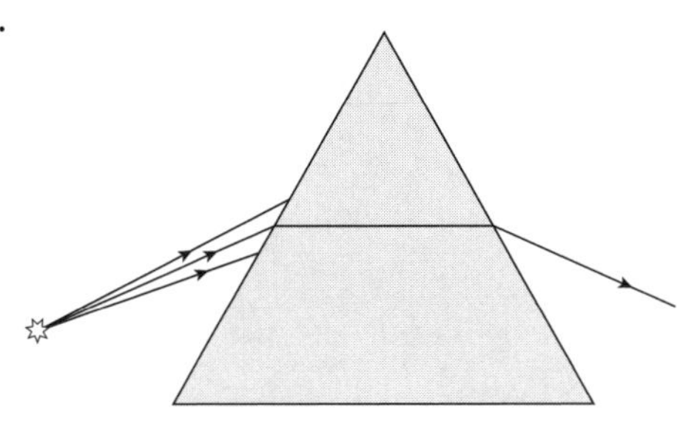

b.

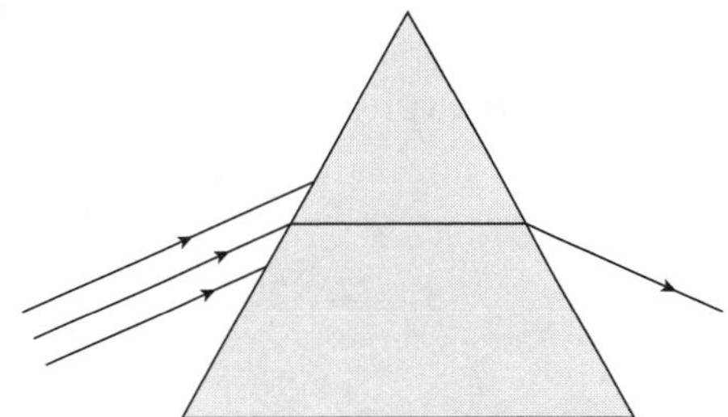

13. Draw the trajectories of seven rays that leave the object heading toward the seven dots on the boundary. Assume $n_2 < n_1$ and $\theta_c = 45°$

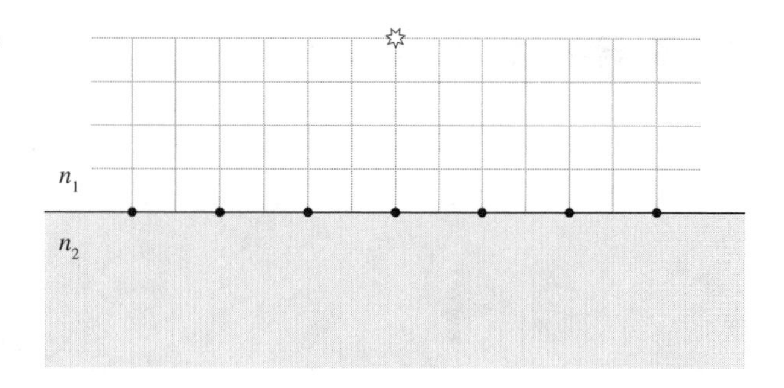

18.4 Color and Dispersion

14. A beam of white light from a flashlight passes through a red piece of plastic.
    a. What is the color of the light that emerges from the plastic? ______
    b. Is the emerging light as intense as, more intense than, or less intense than the white light? Explain.

    c. The light then passes through a blue piece of plastic. Describe the color and intensity of the light that emerges.

15. Suppose you looked at the sky on a clear day through pieces of red and blue plastic oriented as shown. Describe the color and brightness of the light coming through sections 1, 2, and 3.

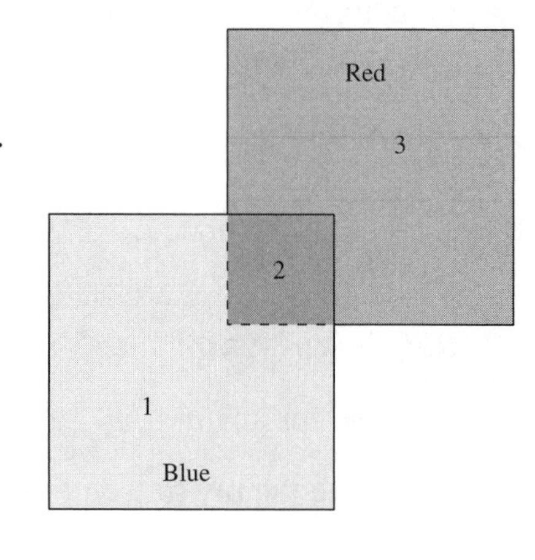

Section 1:

Section 2:

Section 3:

16. Sketch a plausible absorption spectrum for a patch of bright red paint.

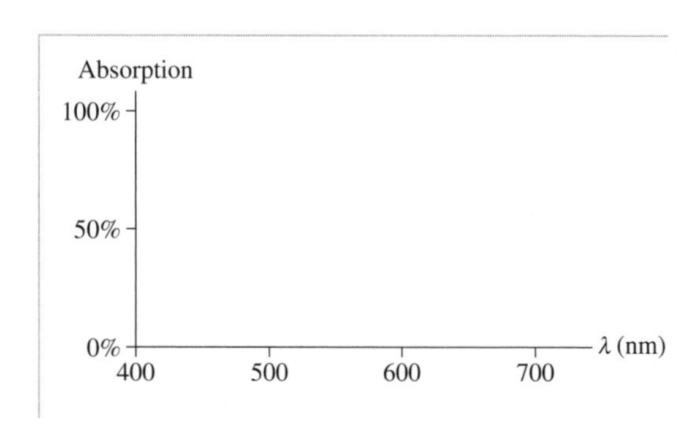

18.5 Image Formation by Refraction

17. a. Draw rays that refract after passing through points B, C, and D. Assume $n_2 > n_1$.

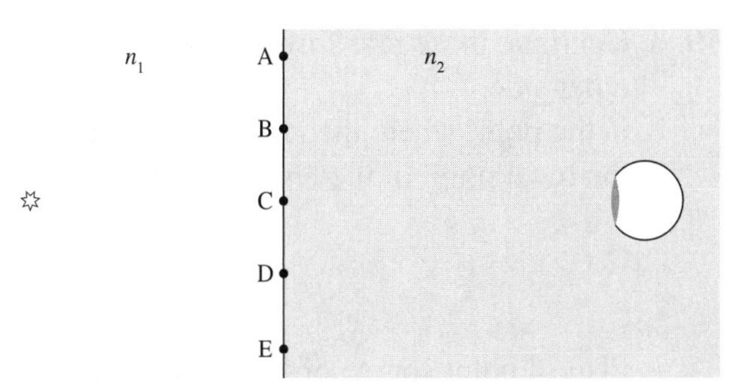

b. Use dotted lines to extend these rays backward into medium 1. Locate and label the image point.

c. Now draw the rays that refract at A and E.

d. Use a different color pen or pencil to draw three rays from the object that enter the eye.

e. Does the distance to the object *appear* to be larger than, smaller than, or the same as the true distance? Explain.

18. A thermometer is partially submerged in an aquarium. The underwater part of the thermometer is not shown.

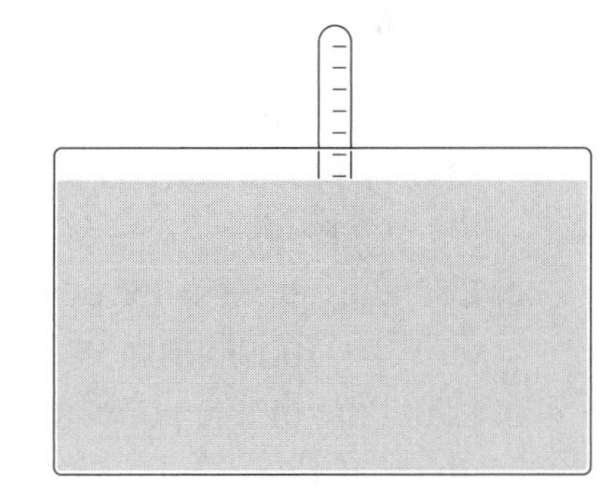

a. As you look at the thermometer, does the underwater part appear to be closer than, farther than, or the same distance as the top of the thermometer?

b. Complete the drawing by drawing the bottom of the thermometer as you think it would look.

18.6 Thin Lenses: Ray Tracing

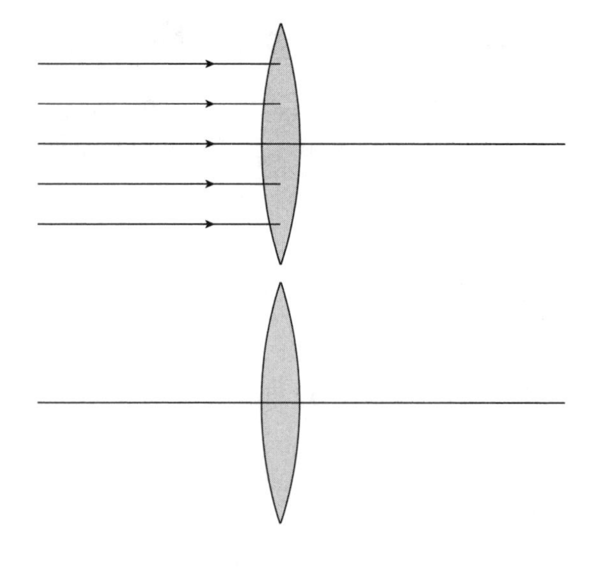

19. a. Continue these rays through the lens and out the other side.
    b. Is the point where the rays converge the same as the focal point of the lens? Or different? Explain.

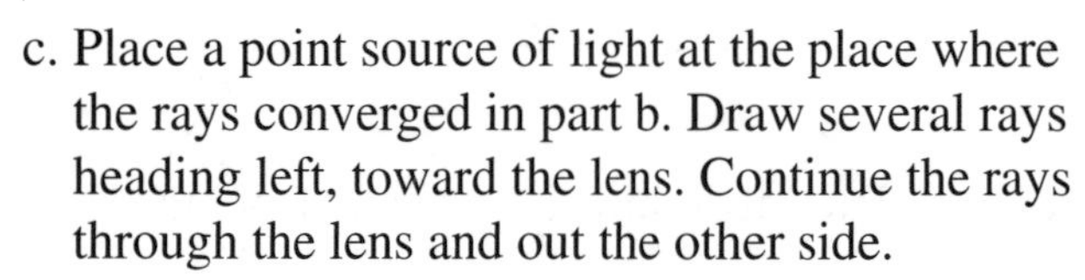

    c. Place a point source of light at the place where the rays converged in part b. Draw several rays heading left, toward the lens. Continue the rays through the lens and out the other side.
    d. Do these rays converge? If so, where?

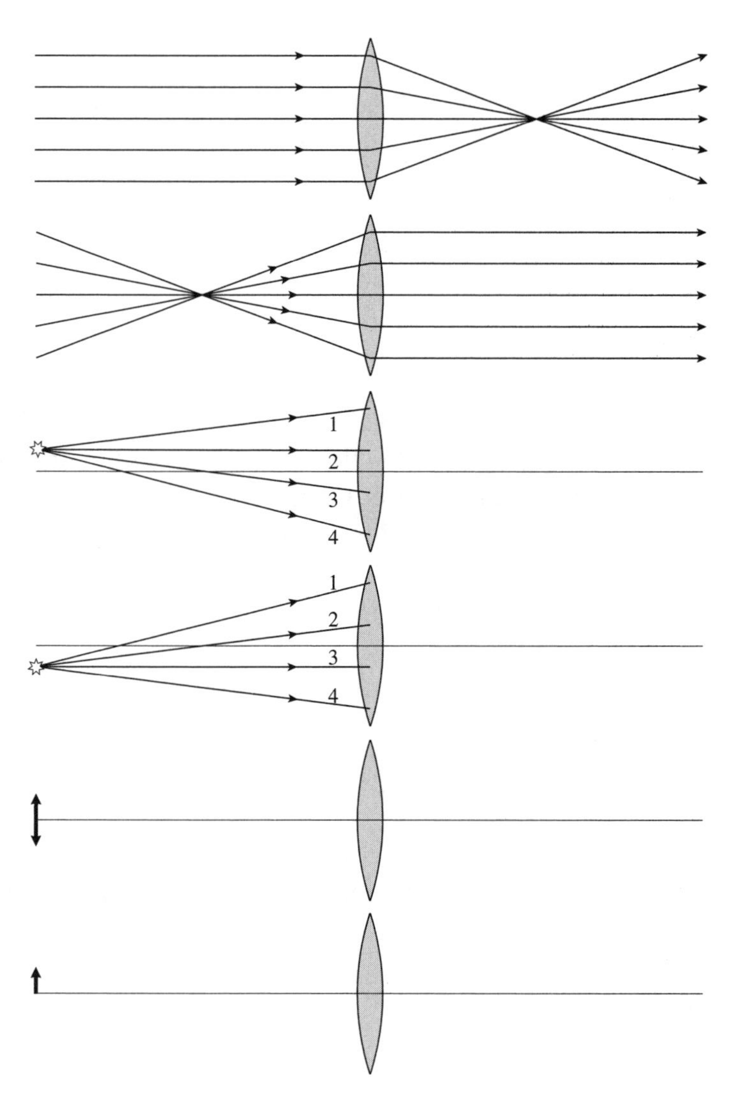

20. The top two figures show test data for a lens. The third figure shows a point source near this lens and four rays heading toward the lens.
    a. For which of these rays do you know, from the test data, its direction after passing through the lens?
    b. Draw the rays you identified in part a as they pass through the lens and out the other side.
    c. Use a different color pen or pencil to draw the trajectories of the other rays.
    d. Label the image point. What kind of image is this?
    e. The fourth figure shows a second point source. Use ray tracing to locate its image point.
    f. The fifth figure shows an extended object. Have you learned enough to locate its image? If so, draw it.
    g. The last figure shows another extended object. Draw its image.

21. An object is near a lens whose focal points are marked with dots.

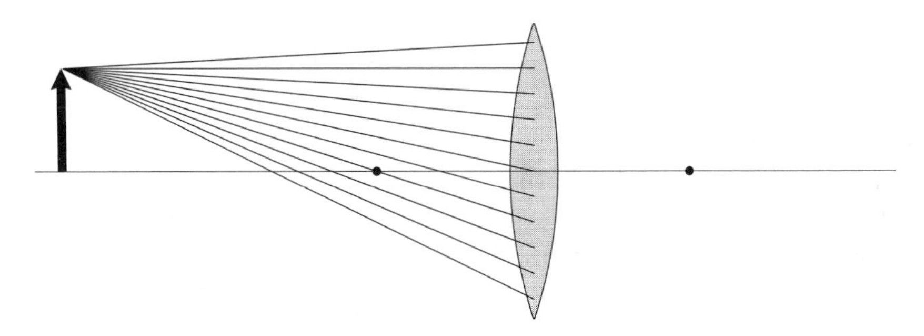

a. Identify the three special rays and continue them through the lens.

b. Use a different color pen or pencil to draw the trajectories of the other rays.

22. An object is near a lens whose focal points are shown.

a. Use ray tracing to locate the image of this object.

b. Is the image upright or inverted? ____________

c. Is the image height larger or smaller than the object height? ____________

d. Is this a real or a virtual image? Explain how you can tell.

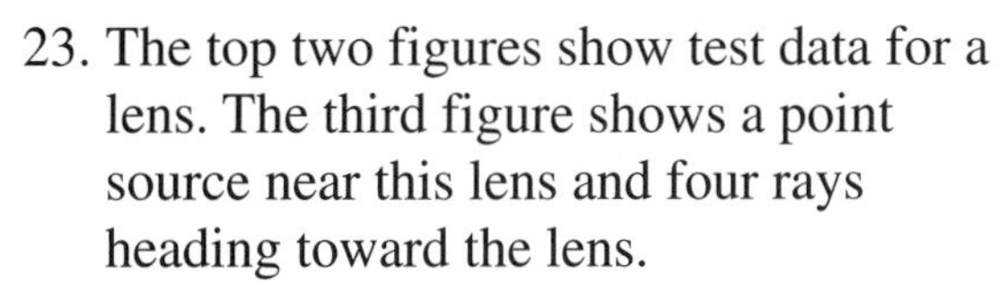

23. The top two figures show test data for a lens. The third figure shows a point source near this lens and four rays heading toward the lens.

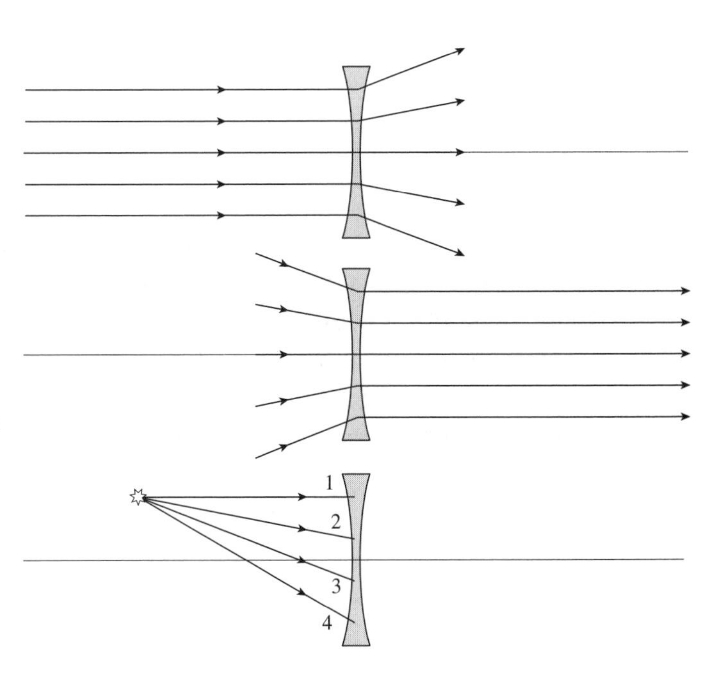

a. For which of these rays do you know, from the test data, its direction after passing through the lens?

b. Draw the rays you identified in part a as they pass through the lens and out the other side.

c. Use a different color pen or pencil to draw the trajectories of the other rays.

d. Find and label the image point. What kind of image is this?

18.7 Image Formation with Spherical Mirrors

24. A 3.0-cm-high object is placed 10.0 cm in front of a concave mirror with a focal length of 4.0 cm. Use ray tracing to determine the location of the image, the orientation of the image, and the height of the image.
    - Locate the mirror on the optical axis shown.
    - Represent the object with an upright arrow at distance 10.0 cm from the mirror.
    - Draw the appropriate three "special rays" to locate the image.

25. A 3.0-cm-high object is placed 5.0 cm in front of a convex mirror with a focal length of 6.0 cm. Use ray tracing to determine the location of the image, the orientation of the image, and the height of the image.
    - Locate the mirror on the optical axis shown.
    - Represent the object with an upright arrow at distance 5.0 cm from the mirror.
    - Draw the appropriate three "special rays" to locate the image.

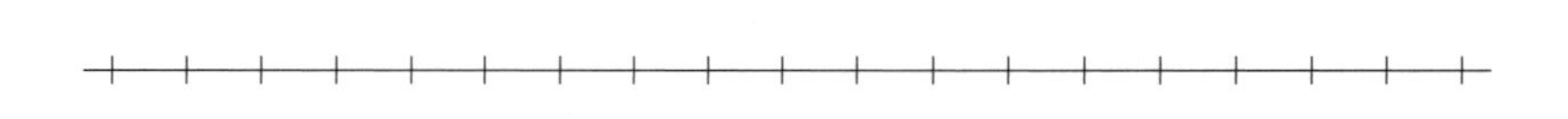

26. A 3.0-cm-high object is placed 10.0 cm in front of a convex mirror with a focal length of −4.0 cm. Use ray tracing to determine the location of the image, the orientation of the image, and the height of the image.
    - Locate the mirror on the optical axis shown.
    - Represent the object with an upright arrow at distance 10.0 cm from the mirror.
    - Draw the appropriate three ‘special rays’ to locate the image.

19 Optical Instruments

19.1 Finding the Image of a Lens or Mirror

1. Consider the thin lens equation: $\frac{1}{s} + \frac{1}{s'} = \frac{1}{f}$

   a. i. If s, s', and f are all positive, can either s or s' ever be smaller than f? If so, give an example. If not, why not? What does your answer imply about the location of a real image?

   ii. If s, s', and f are all positive, for a given value of f, how will s' change as s is increased? How will s' change as s is decreased? What does your answer imply about the size, orientation, and location of the image compared to the object?

   b. i. If $f > s > 0$, can s' ever be positive? If so, give an example. If not, why not? What does your answer imply about the location of the image?

   ii. If $f > s > 0$, can $|s'|$ ever be less than s? If so, give an example. If not, why not? What does your answer imply about the size, orientation, and location of the image?

   c. If $f < 0$ and $s > 0$ can $|s'|$ ever be greater than s? If so, give an example. If not, why not? What does your answer imply about the size, orientation, and location of the image?

2. Two rays are shown to locate the image of the arrow below.
   a. On the ray diagram, distinctly mark and label three pairs of similar right triangles.

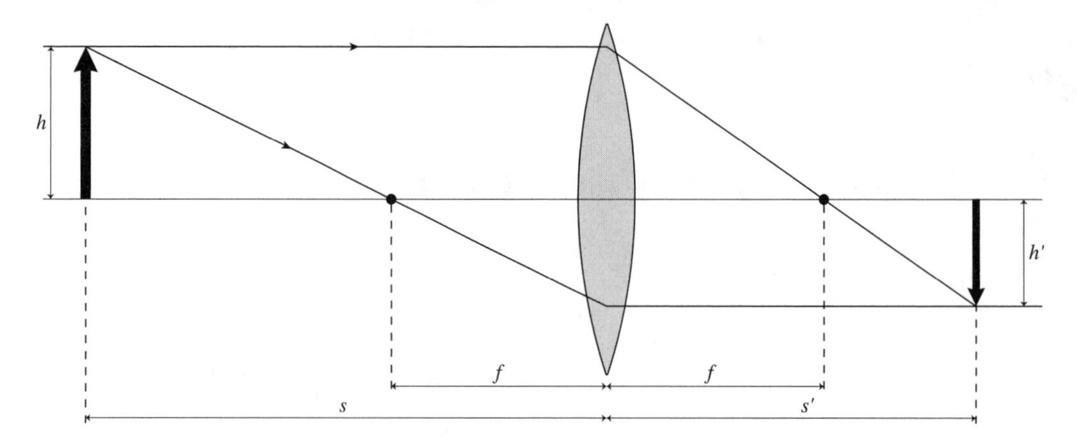

   b. When two triangles are similar, then the ratio of the lengths of any two sides of one triangle are the same as the ratio of the lengths of the corresponding sides of the other triangle. Using the quantities given on the diagram, write three equations that compare the respective ratios of two sides from the pairs of similar right triangles that you have marked above. (Note that though any two sides can be used, the hypotenuse is generally not a convenient choice for this comparison.)

3. The object and final image formed by a pair of lenses is shown. Complete the diagram by drawing three principal rays and indicating the location of the intermediate image.

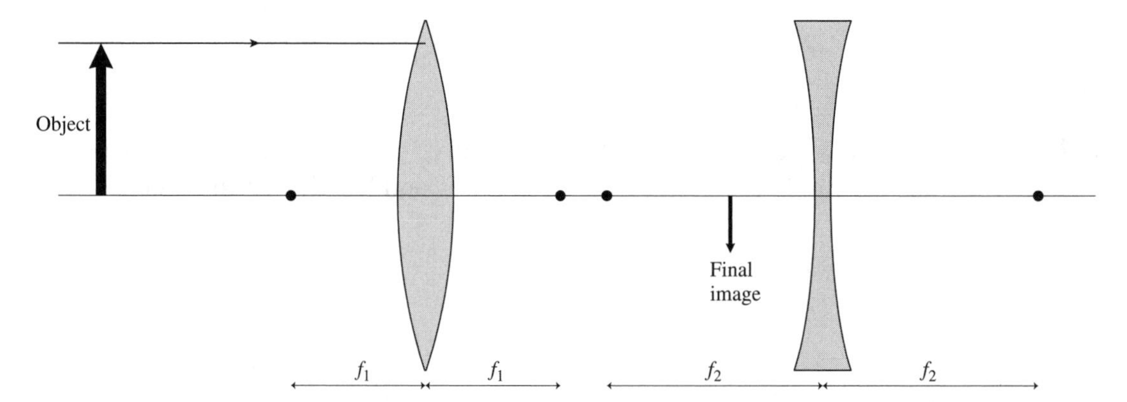

4. Each of the lenses shown is curved only on one surface and with differing amounts of curvature.

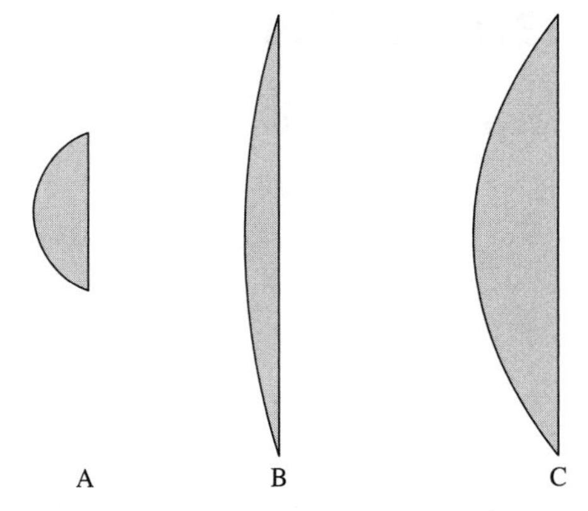

   a. Rank, in order from longest to shortest, the focal lengths of the lenses.

   Order:

   Explanation:

   b. Rank, in order from greatest to least, the powers of the lenses.

   Order:

   Explanation:

19.2 The Camera

5. A camera iris diaphragm set to $f/4$ is shown.

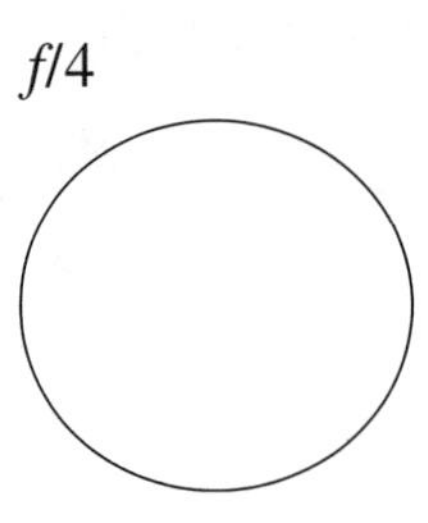

a. In the spaces provided, redraw the iris opening for the given f-stop if the same lens is used:

i. $f/8$

ii. $f/2.8$.

b. Assuming equal illumination is required in each case, rank the diaphragm settings above, from greatest to least, in terms of the shutter speed required. Note that a greater shutter speed corresponds to a smaller amount of time exposure for the image.

Order:

Explanation:

19.3 The Human Eye

6. Though the human eye and a camera have many similarities, one fundamental difference is how each is typically adjusted to create an in-focus image for objects at different distances. In a camera, the image is formed on the film by the camera's lens. At the back of your eye, the image is formed on the retina by the curvature of the cornea and the lens of the eye.
   a. What simple camera adjustment enables the image on the film to remain in focus as the object is moved away from the camera? State both the parameter to be adjusted and whether it is increased or decreased.

   b. What eye adjustment is required to keep the image on the back of your retina in focus as the object you are looking at is moved away from your eye? State both the parameter to be adjusted and whether it is increased or decreased.

   c. Why are the methods for adjusting the camera and the eye for different object distances not more similar? What prevents the camera from being adjusted in the manner of your eye? Why can't the eye be adjusted for different object distances in the same manner as the camera?

7. The figures show rays of light entering a model of the eye in a failed attempt to form an image. As can be seen, the image in each case has been formed either in front of or behind the retina.
   a. Identify which drawing corresponds to a hyperopic eye and which corresponds to a myopic eye.
   b. Redraw the eye in the space at right with an appropriate corrective lens just in front of the eye. Show how rays passing through this lens will be altered to form an image on the retina.

   a. hyperopic or myopic ____________

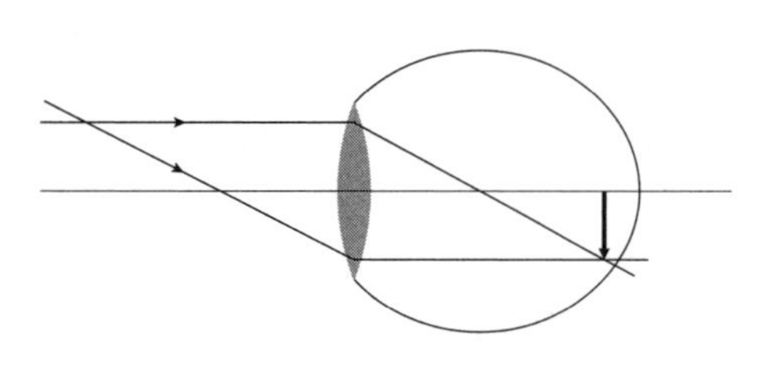

   b. hyperopic or myopic ____________

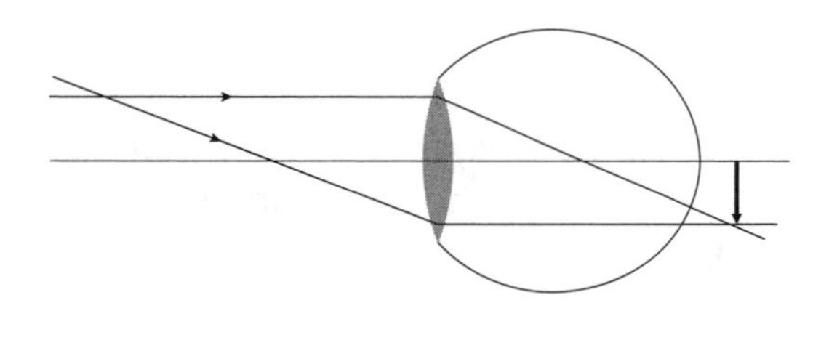

8. If the image from a diverging lens is always minified compared to the original object, how can a diverging lens help someone see objects that are far away more clearly? Use a ray diagram to justify your answer.

19.4 The Magnifier

9. a. Draw a ray diagram below to show the location of the image produced by this lens.

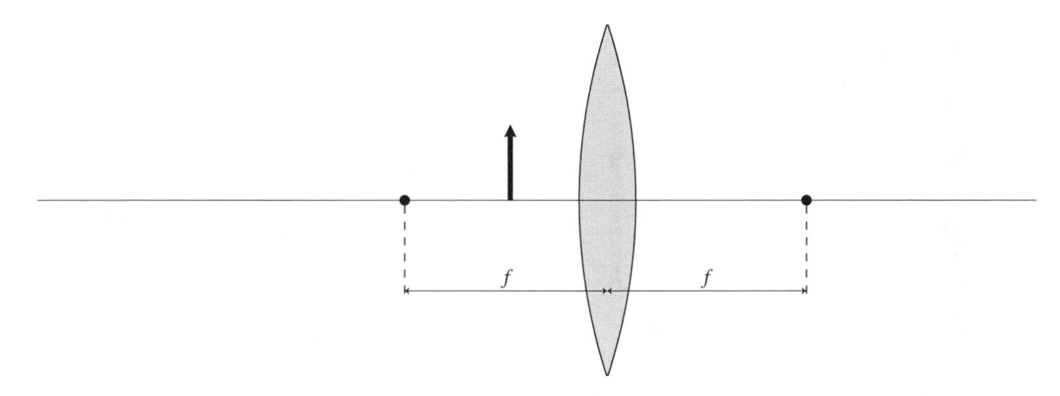

b. On the diagram above, indicate the range of locations from which a viewer could see the image formed by the lens, i.e., indicate where the viewer's eye would have to be located to see the image.

c. Simple magnifiers are sometimes misleadingly said to make the object appear "closer" so that you can see it better. Is the image of the object in fact closer to the observer than the object? Explain.

d. Given that the normal near point of the human eye is 25 cm, can a simple magnifier enable you to see objects that are closer than 25 cm away? Explain.

10. The figure below shows the formation of an image on the retina due to the two arrows, A and B, shown. The eye adjusts in each case to form an in-focus image on the back of the retina.

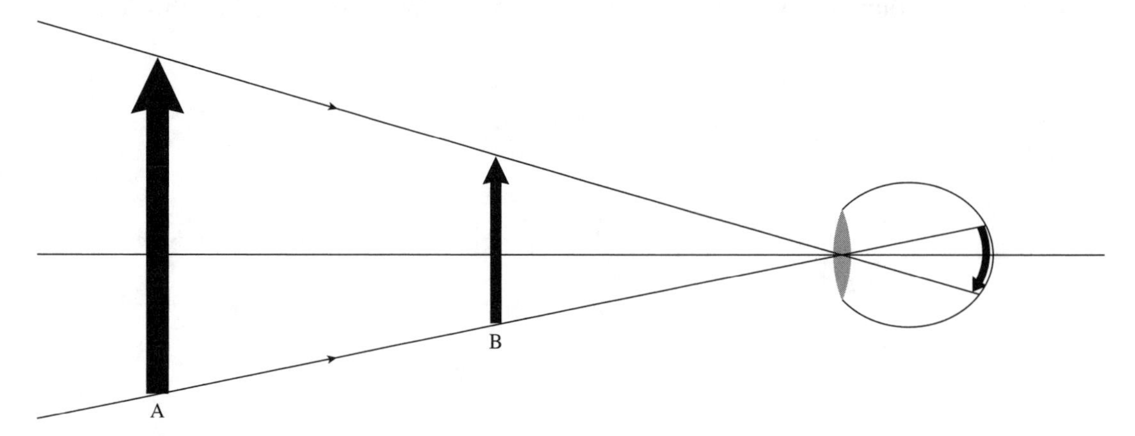

a. For which of these arrows is the magnitude of the magnification of the image formed by the eye greater? Explain.

b. Is the apparent angular size of one of these images greater than the other? If so, which one is greater? Explain.

19.5 The Microscope

11. a. Complete the ray diagram by drawing two principal rays to show how the eyepiece affects light from the intermediate image of the objective. Show that the two rays are parallel on the right side of the eyepiece. (Because these rays are parallel, it is not possible to draw the final virtual image on your diagram.) Indicate on the diagram a location along the center axis from which an observer could see the final image.

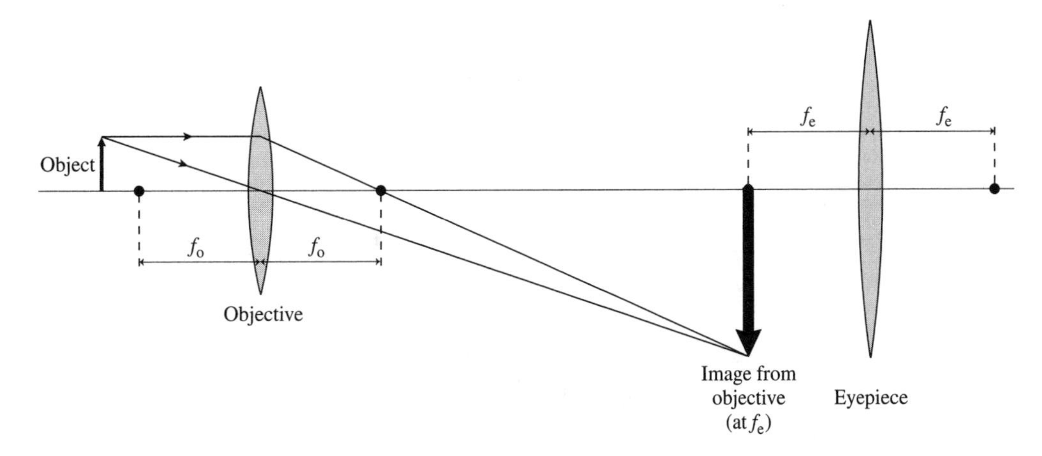

b. On your diagram above, indicate the angle subtended by the final image. This is the image's angular size. Measure this angle with a protractor.
θ final image = ______________

c. Use a red pencil to draw a line from the location of the observer to the tip of the object. Indicate the angular size of the original object and measure this angle with a protractor.
θ object = ______________

d. How does the angular size of the object compare to the angular size of the final image? What is the magnification of the two-lens system?

19.6 The Telescope

12. Why does the magnification of a telescope vary directly as the focal length of the objective, but the magnification of a microscope varies inversely as the objective's focal length? [Hint: Think about the location and size of the objects seen with a microscope and telescope, respectively.]

13. The figure indicates the relative sizes and locations of the lenses for three simple telescopes. Rank, in order from greatest to least, the magnifications of the telescopes M_1, M_2, and M_3.

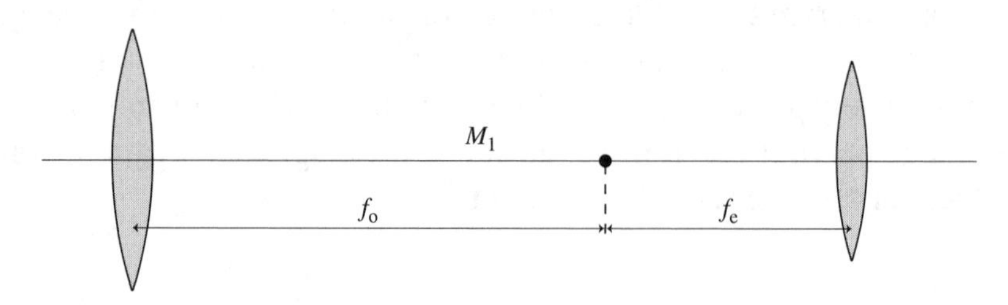

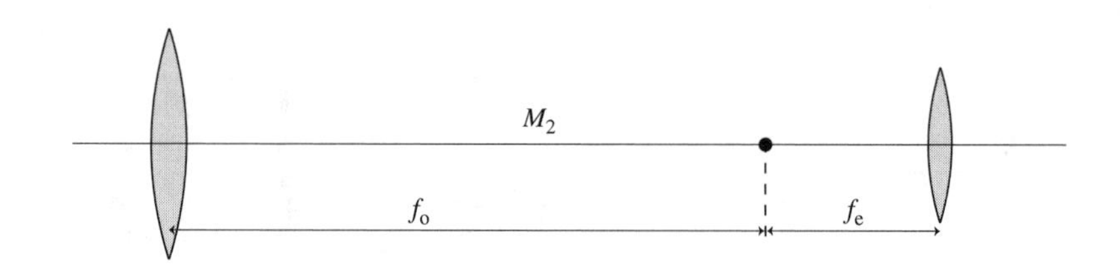

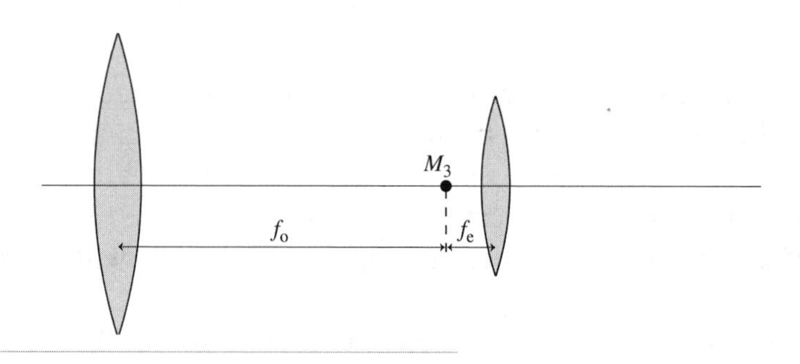

Order:

Explanation:

14. Though the eyepieces for microscopes and telescopes are similar in diameter in order to be comfortable for viewing, the objective lens of a telescope is typically much larger in diameter than the eyepiece, while the objective lens for a microscope is typically smaller than the eyepiece. Why does a good telescope use a very large diameter objective, whereas a good microscope does not?

19.7 Resolution of Optical Instruments

15. Parallel rays are incident on the lens at various distances from the center of the lens. Trace the rays through the lens to show how the rays that pass nearer to the center focus further from the lens than those that pass near its edges. The spreading of the rays intersection points should show the effects of spherical aberration.

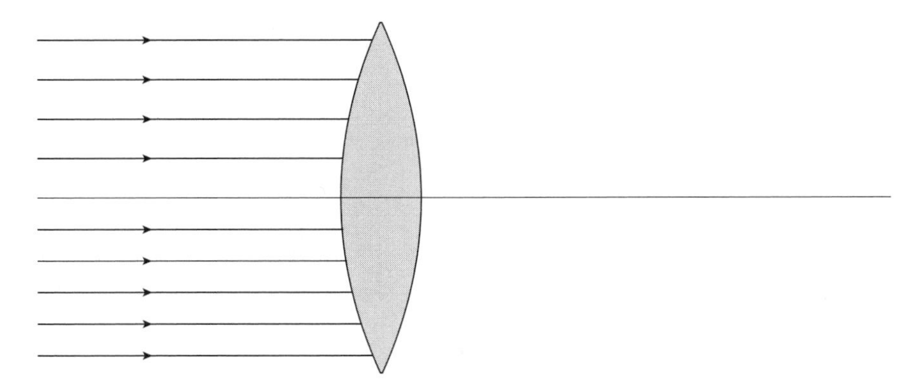

16. Draw and label blue and red rays to show the effects of chromatic aberration on the divergence of the rays after passing through the lens. Assume $n_{blue} > n_{red}$.

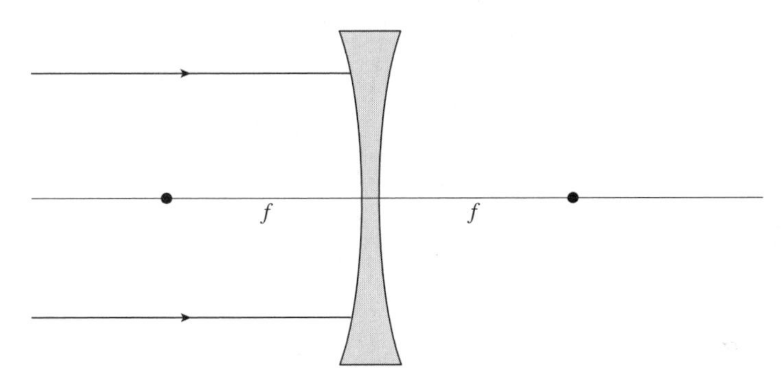

17. Lenses and curved mirrors can both exhibit the effects of spherical aberration. Can mirrors and lenses both also exhibit chromatic aberration? Explain.

18. Just the central maxima of the overlapping diffraction patterns are shown for six sources. For each overlapping pair, sketch the sum of the intensity of the two sources in the pair directly on top of the pattern shown, using the same intensity scale.

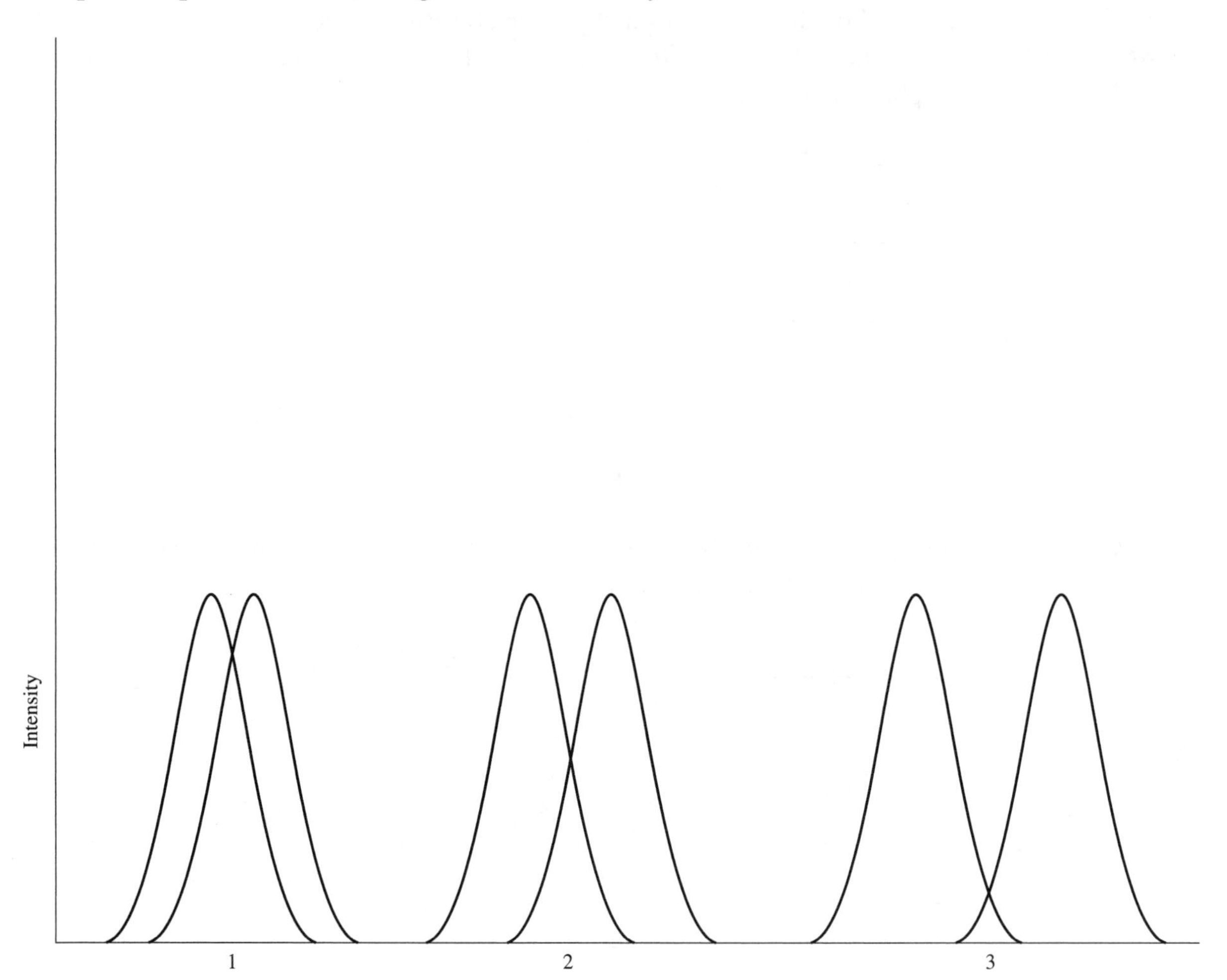

For which of these pairs could the two sources be distinguished when only the combined intensities can be detected? Explain.

20 Electric Forces and Fields

20.1 Charges and Forces

1. Two lightweight balls hang straight down when both are neutral. They are close enough together to interact, but not close enough to touch. Draw pictures showing how the balls hang if:

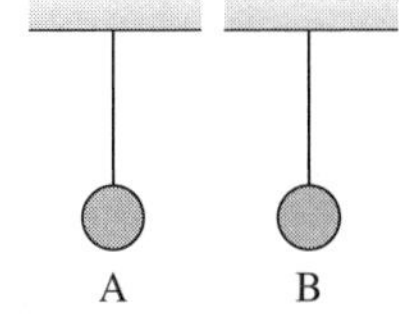

   a. Both are touched with a plastic rod that was rubbed with wool.

   b. The two charged balls of part a are moved farther apart.

   c. Ball A is touched by a plastic rod that was rubbed with wool and ball B is touched by a glass rod that was rubbed with silk.

   d. Both are charged by a plastic rod, but ball A is charged more than ball B.

   e. Ball A is charged by a plastic rod. Ball B is neutral.

   f. Ball A is charged by a glass rod. Ball B is neutral.

2. After combing your hair briskly, the comb will pick up small pieces of paper.
   a. Is the comb charged? Explain.

   b. How can you be sure that it isn't the paper that is charged? Propose an experiment to test this.

   c. Is your hair charged after being combed? What evidence do you have for your answer?

   d. What kind of charge is the comb likely to have? Why?

   e. How could you test your answer to part d?

3. You've been given a piece of material. Propose an experiment or a series of experiments to determine if the material is a conductor or an insulator. State clearly what the outcome of each experiment will be if the material is a conductor and if it is an insulator.

4. Suppose there exists a third type of charge in addition to the two types we've called glass and plastic. Call this third type X charge. What experiment or series of experiments would you use to test whether an object has X charge? State clearly how each possible outcome of the experiments is to be interpreted.

5. A negatively charged electroscope has separated leaves.

   a. Suppose you bring a negatively charged rod close to the top of the electroscope, but not touching. How will the leaves respond? Use both charge diagrams and words to explain.

   b. How will the leaves respond if you bring a positively charged rod close to the top of the electroscope, but not touching? Use both charge diagrams and words to explain.

6. A lightweight, positively charged ball and a neutral rod hang by threads. They are close but not touching. A positively charged glass rod touches the hanging rod on the end opposite the ball, then the rod is removed.

   Touch

   a. Draw a picture of the final positions of the hanging rod and the ball if the rod is made of glass.

   b. Draw a picture of the final positions of the hanging rod and the ball if the rod is metal.

7. a. Metal sphere A is initially neutral. A positively charged rod is brought near, but not touching. Is A now positive, negative, or neutral? Explain.

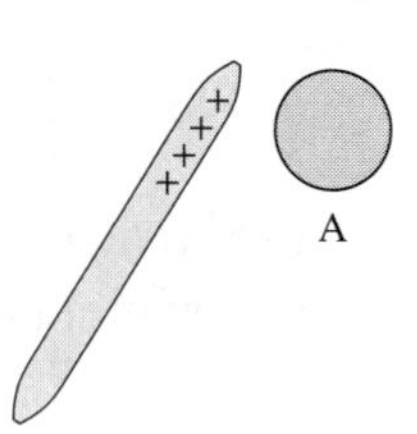

b. Metal spheres A and B are initially neutral and are touching. A positively charged rod is brought near A, but not touching. Is A now positive, negative, or neutral? Explain.

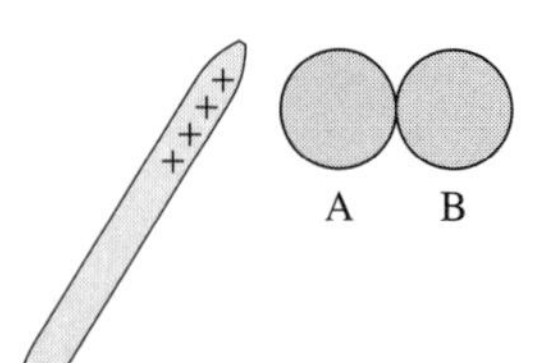

c. Metal sphere A is initially neutral. It is connected by a metal wire to the ground. A positively charged rod is brought near, but not touching. Is A now positive, negative, or neutral? Explain.

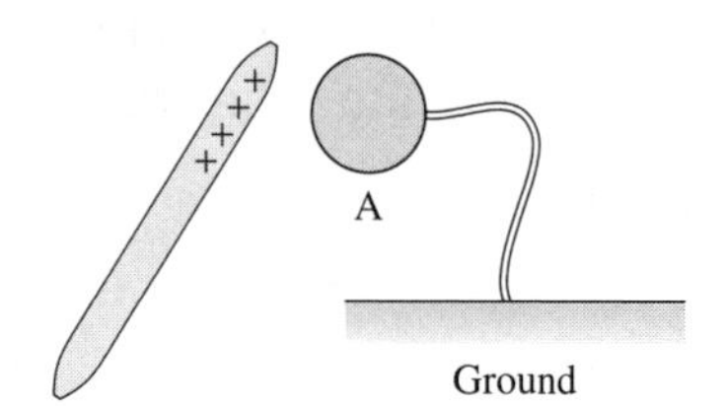

8. A lightweight, positively charged ball and a neutral metal rod hang by threads. They are close but not touching. A positively charged rod is held close to, but not touching, the hanging rod on the end opposite the ball.

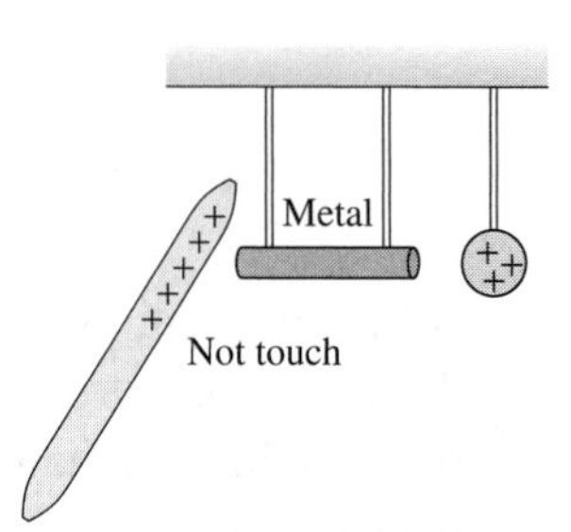

a. Draw a picture of the final positions of the hanging rod and the ball. Explain your reasoning.

b. Suppose the positively charged rod is replaced with a negatively charged rod. Draw a picture of the final positions of the hanging rod and the ball. Explain your reasoning.

20.2 Charges, Atoms, and Molecules

9. Two oppositely charged metal spheres have equal quantities of charge. They are brought into contact with a neutral metal rod.

   a. What is the final charge state of each sphere and of the rod?

   b. Give a microscopic explanation, in terms of fundamental charges, of how these final states are reached. Use both charge diagrams and words.

10. a. A negatively charged plastic rod touches a neutral piece of metal. What is the final charge state (positive, negative, or neutral) of the metal? Use both charge diagrams and words to explain how this charge state is achieved.

    b. A positively charged glass rod touches a neutral piece of metal. What is the final charge state of the metal? Use both charge diagrams and words to explain how this charge state is achieved.

11. If you bring your finger near a lightweight, negatively charged hanging ball, the ball swings over toward your finger. Use charge diagrams and words to explain this observation.

Finger

12. The figure shows an atom with four protons in the nucleus and four electrons in the electron cloud.

    a. Draw a picture showing how this atom will look if a positive charge is held just *above* the atom.

    b. Is there a net force on the atom? If so, in which direction? Explain.

20.3 Coulomb's Law

13. For each pair of charges, draw a force vector *on each charge* to show the electric force acting on that charge. The length of each vector should be proportional to the magnitude of the force. Each + and – symbol represents the same quantity of charge.

a.

b.

c.

d.

14. For each group of charges, use a **black** pen or pencil to draw the forces acting on the gray positive charge. Then use a **red** pen or pencil to show the net force on the gray charge. Label $\vec{F}_{net}$.

a.

b.

c.

d.

e.

f.

15. Can you assign charges (positive or negative) so that these forces are correct? If so, show the charges on the figure. (There may be more than one correct response.) If not, why not?

a.

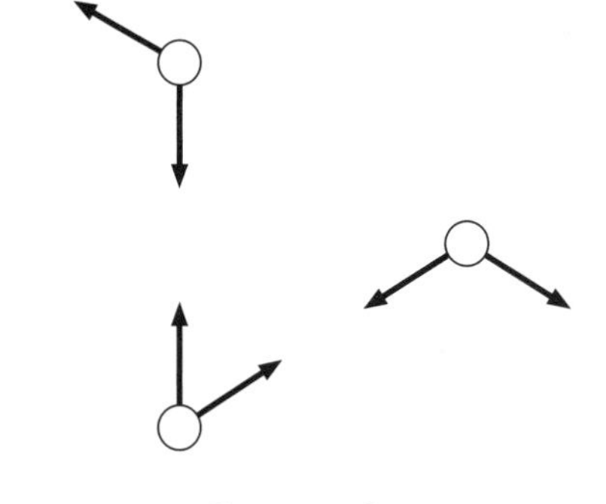

b.

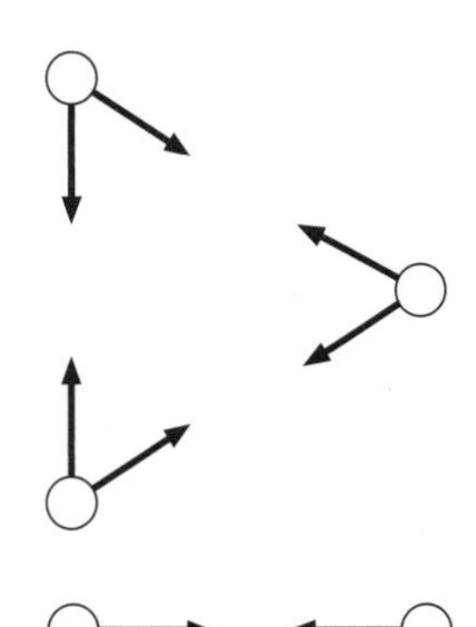

c.

d.

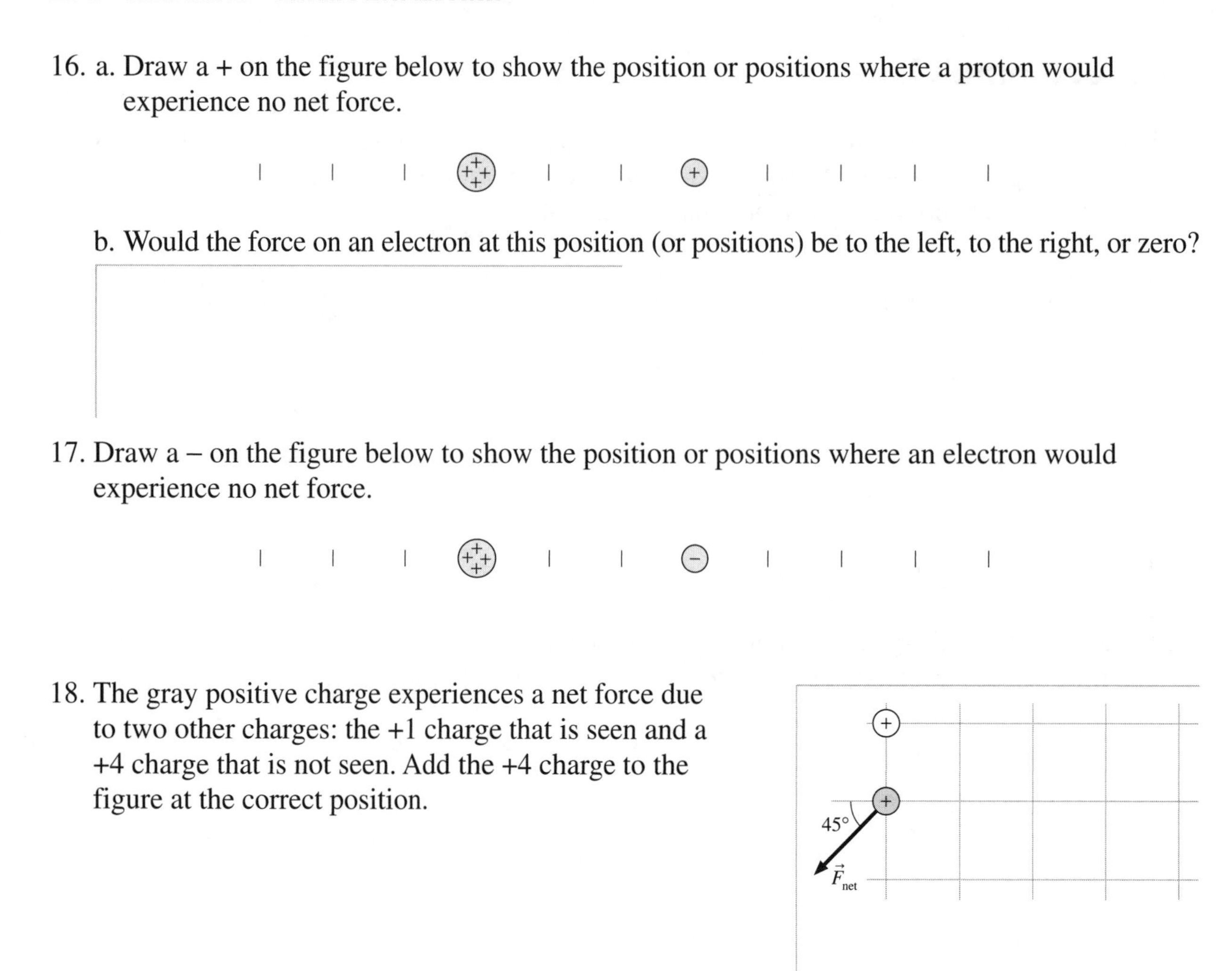

16. a. Draw a + on the figure below to show the position or positions where a proton would experience no net force.

b. Would the force on an electron at this position (or positions) be to the left, to the right, or zero?

17. Draw a – on the figure below to show the position or positions where an electron would experience no net force.

18. The gray positive charge experiences a net force due to two other charges: the +1 charge that is seen and a +4 charge that is not seen. Add the +4 charge to the figure at the correct position.

19. In your own words, describe what is meant by a "point charge."

20.4 The Concept of the Electric Field

20. This is a uniform gravitational field near the earth's surface. Rank in order, from largest to smallest, the accelerations a_1 to a_3 of a small mass at points 1, 2, and 3.

Order:

Explanation:

21. This is the gravitational field of the earth. Rank in order, from largest to smallest, the accelerations a_1 to a_3 of a small mass at points 1, 2, and 3.

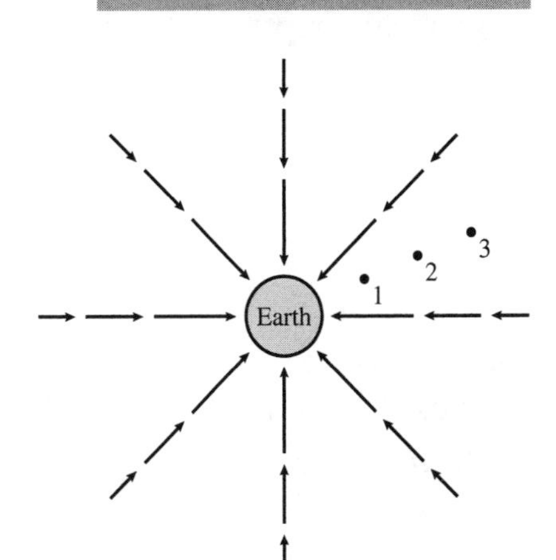

Order:

Explanation:

22. At points 1 to 4, draw an electric field vector with the proper direction and whose length is proportional to the electric field strength at that point.

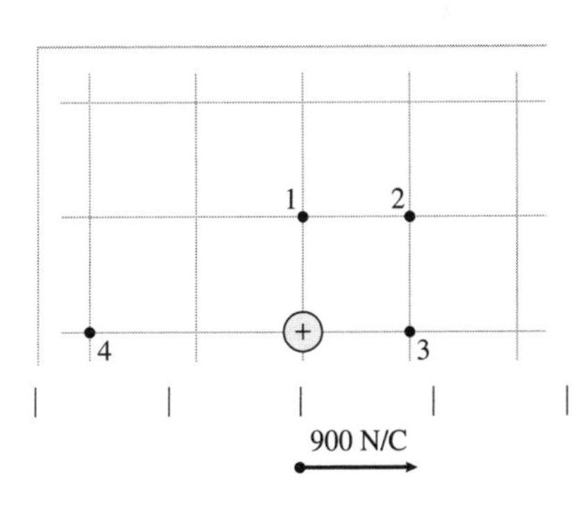

23. a. The electric field of a point charge is shown at *one* point in space. Can you tell if the charge is + or –? If not, why not?

b. Here the electric field of a point charge is shown at two positions in space. Now can you tell if the charge is + or –? Explain.

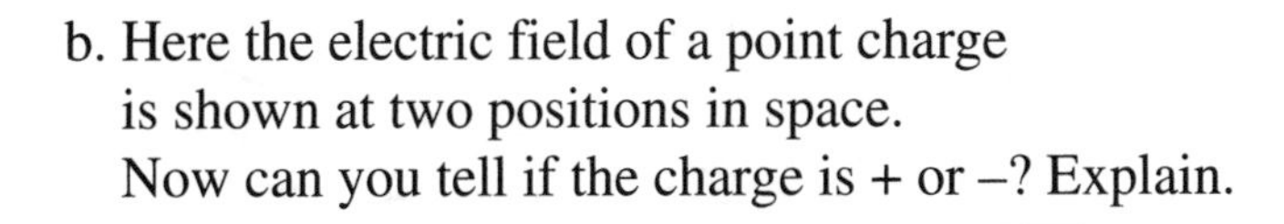

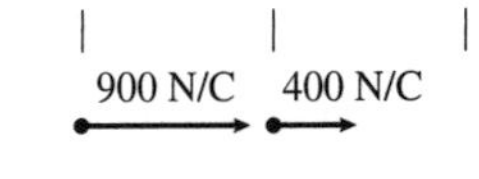

c. Can you determine the location of the charge? If so, draw it on the figure. If not, why not?

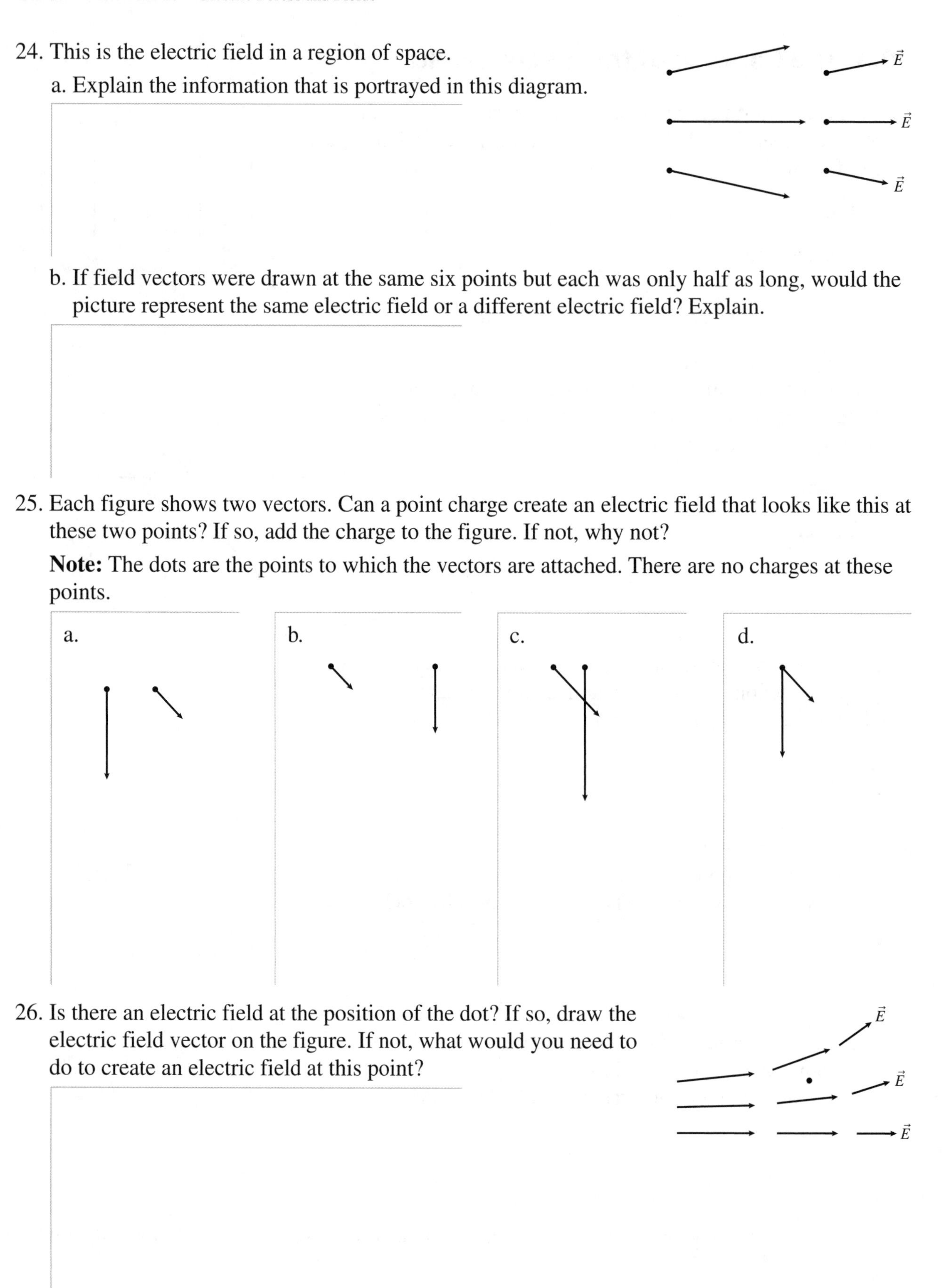

24. This is the electric field in a region of space.

a. Explain the information that is portrayed in this diagram.

b. If field vectors were drawn at the same six points but each was only half as long, would the picture represent the same electric field or a different electric field? Explain.

25. Each figure shows two vectors. Can a point charge create an electric field that looks like this at these two points? If so, add the charge to the figure. If not, why not?

Note: The dots are the points to which the vectors are attached. There are no charges at these points.

26. Is there an electric field at the position of the dot? If so, draw the electric field vector on the figure. If not, what would you need to do to create an electric field at this point?

20.5 Applications of the Electric Field

27. At each of the dots, use a **black** pen or pencil to draw and label the electric fields $\vec{E}_1$ and $\vec{E}_2$ due to the two point charges. Make sure that the *relative* lengths of your vectors indicate the strength of each electric field. Then use a **red** pen or pencil to draw and label the net electric field $\vec{E}_{net}$.

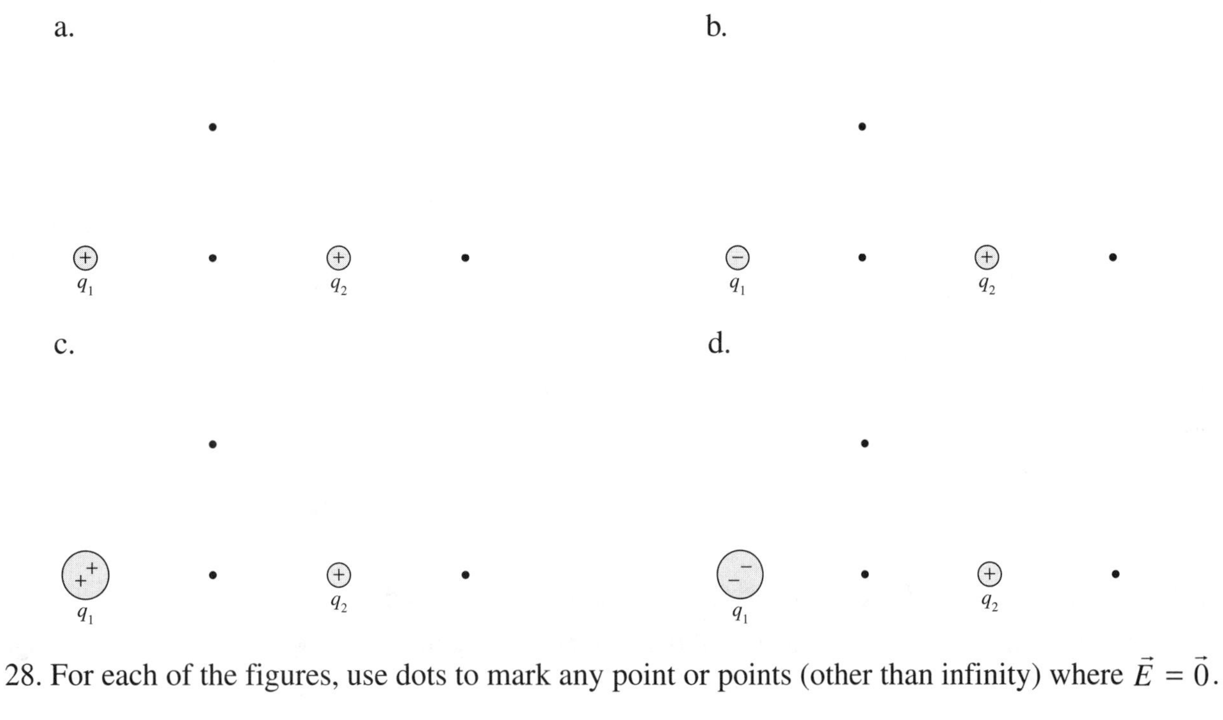

28. For each of the figures, use dots to mark any point or points (other than infinity) where $\vec{E} = \vec{0}$.

a.

b.

29. Use a **black** pen or pencil to draw the two electric fields $\vec{E}_1$ and $\vec{E}_2$ at each dot. Then use a **red** pen or pencil to draw $\vec{E}_{net}$. The lengths of your vectors should indicate the magnitude of $\vec{E}$ at each point.

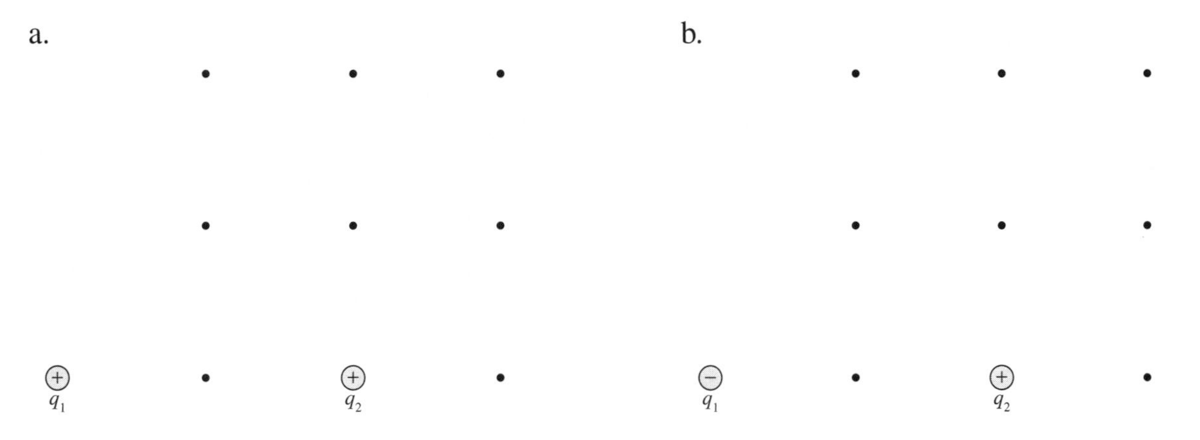

30. For each figure, draw and label the net electric field vector $\vec{E}_{net}$ at each of the points marked with a dot or, if appropriate, label the dot $\vec{E}_{net} = \vec{0}$. The lengths of your vectors should indicate the magnitude of $\vec{E}$ at each point.

a.

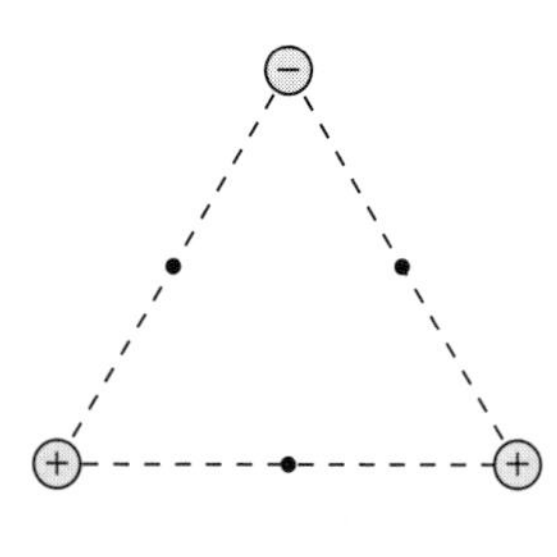

b.

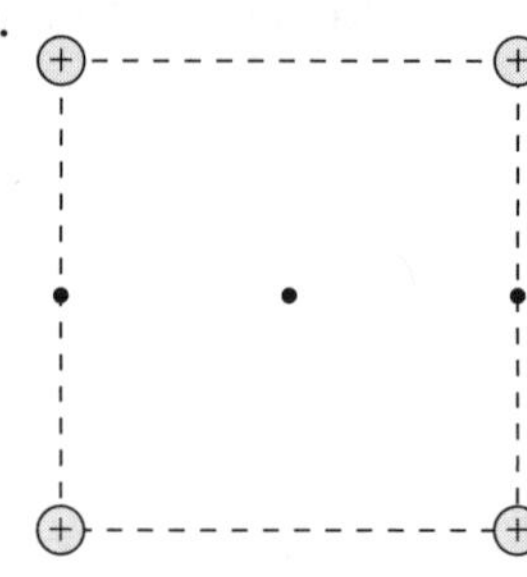

c.

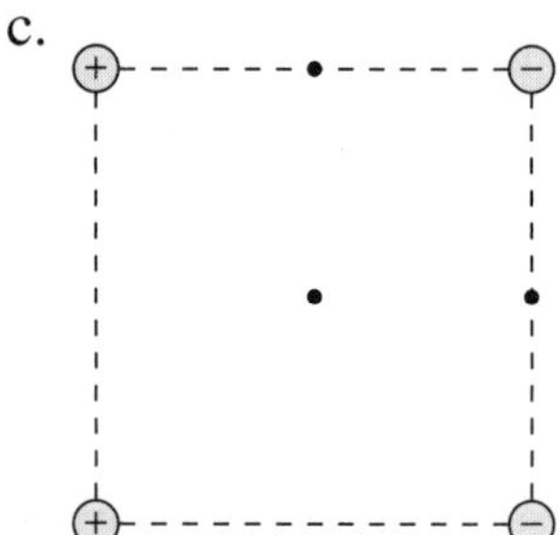

d.

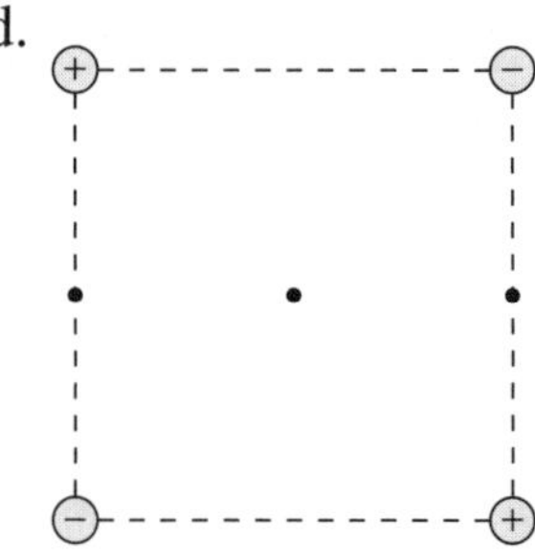

31. Draw the electric field vector at the three points marked with a dot.
Hint: Think of the charges as horizontal positive/negative pairs, then use superposition.

32. Rank in order, from largest to smallest, the electric field strengths E_1 to E_5 at each of these points.

Order:

Explanation:

33. A parallel-plate capacitor is constructed of two square plates, size $L \times L$, separated by distance d. The plates are given charge $\pm Q$. What is the ratio E_f/E_i of the final electric field strength E_f to the initial electric field strength E_i if:

    a. Q is doubled?

    b. L is doubled?

    c. d is doubled?

34. The figure shows the electic field lines in a region of space. Draw the electric field vectors at the three dots.

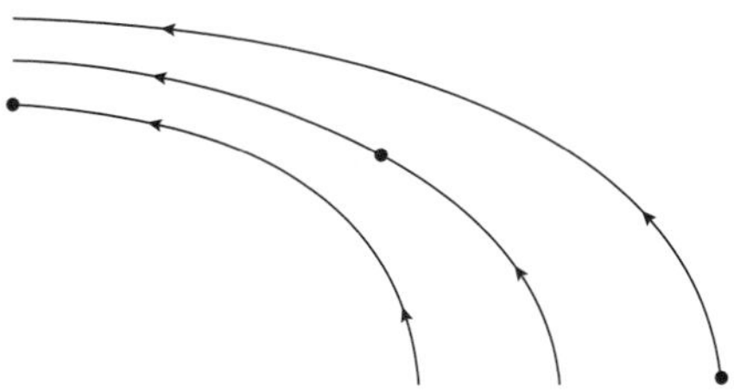

20.6 Conductors in Electric Fields

35. A neutral metal rod is suspended in the center of a parallel-plate capacitor. Then the capacitor is charged as shown.

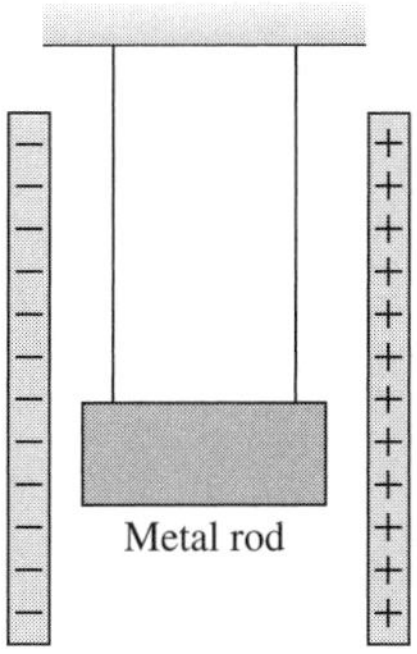

    Metal rod

    a. Is the rod now positive, negative, or neutral? Explain.

    b. Is the rod polarized? If so, draw plusses and minuses on the figure to show the charge distribution. If not, why not?

    c. Does the rod swing toward one of the plates, or does it remain in the center? If it swings, which way? Explain.

36. An insulating thread is used to lower a positively charged metal ball into a metal container. Initially, the container has no net charge. Use plus and minus signs to show the charge distribution on the inner and outer surfaces of the container and any charge on the ball. (The ball's charge is already shown in the first frame.)

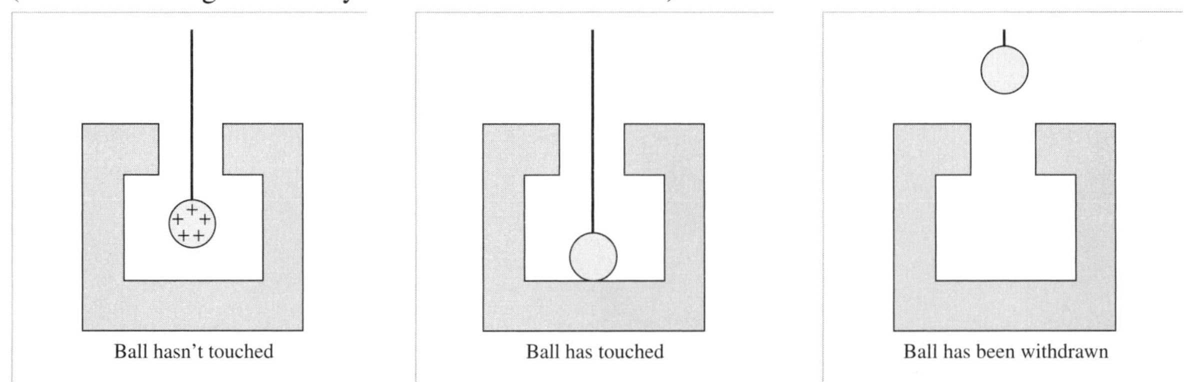

20.7 Forces and Torques on Charges in Electric Fields

37. Positively and negatively charged objects, with equal masses and equal quantities of charge, enter the capacitor in the directions shown.
    a. Use solid lines to draw their trajectories on the figure if their initial velocities are fast.
    b. Use dotted lines to draw their trajectories on the figure if their initial velocities are slow.

38. An electron is launched from the positive plate at a 45° angle. It does not have sufficient speed to make it to the negative plate. Draw its trajectory on the figure.

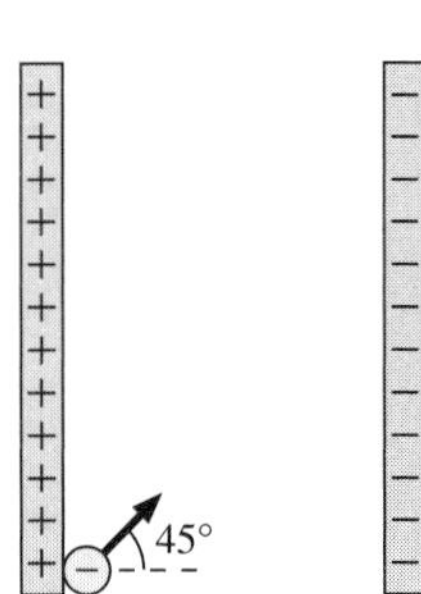

39. The figure shows an electron orbiting a proton in a hydrogen atom.
    a. What force or forces act on the electron?

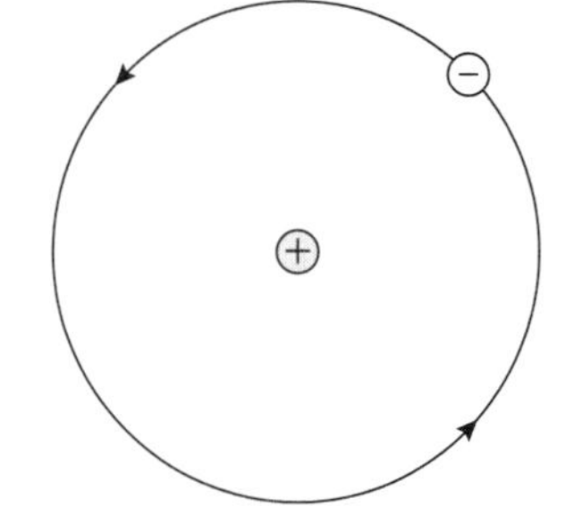

    b. Draw and label the following vectors on the figure: the electron's velocity $\vec{v}$ and acceleration $\vec{a}$, the net force $\vec{F}_{net}$ on the electron, and the electric field $\vec{E}$ at the position of the electron.

21 Electrical Potential

21.1 Electric Potential Energy and the Electric Potential

21.2 Using the Electric Potential

1. Positive and negative point charges are inside a parallel-plate capacitor. The point charges interact only with the capacitor, not with each other.
   a. Use a **black** pen or pencil to draw the electric field vectors inside the capacitor.
   b. Use a **red** pen or pencil to draw the forces acting on the two charges.
   c. Pick a point of your choosing for the zero of potential energy. Label it "$U = 0$" on the diagram.
   d. Is the potential energy of the *positive* point charge positive, negative, or zero? Explain.

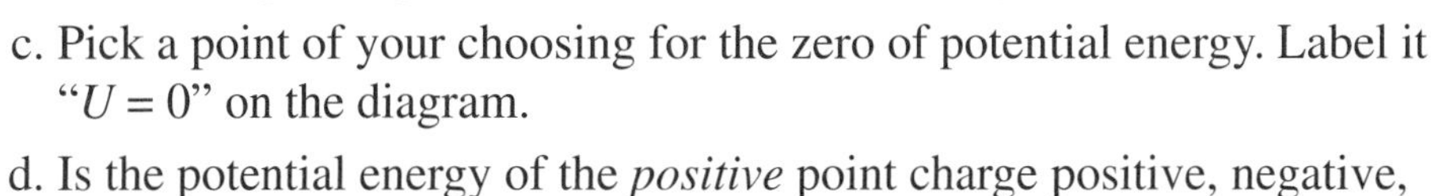

   e. In which direction (right, left, up, or down) does the potential energy of the positive charge decrease? Explain.

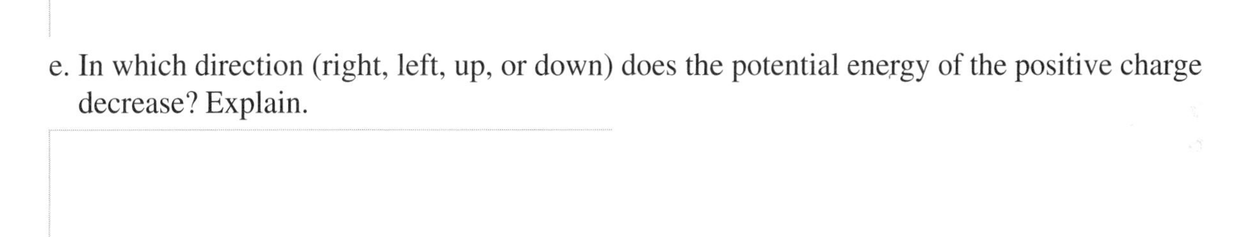

   f. In which direction will the positive charge move if released from rest? Use the concept of energy to explain your answer.

   g. Does your answer to part f agree with the force vector you drew in part b? ____________

   h. Repeat steps d to g for the *negative* point charge.

2. Charged particles with $q = +0.1$ C are fired with 10 J of kinetic energy toward a region of space in which there is an electric potential. The figure shows the kinetic energy of the charged particles as they arrive at nine different points in the region. Determine the electric potential at each of these points. Write the value of the potential above each of the dots. Assume that the particles start from a point where the electric potential is zero.

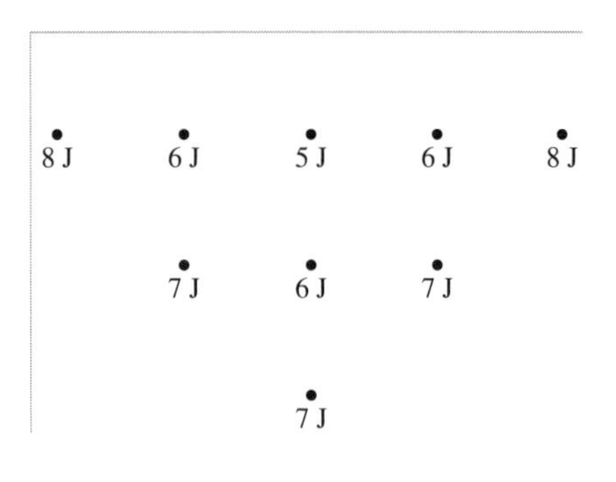

3. Charge q is fired toward a stationary positive point charge.

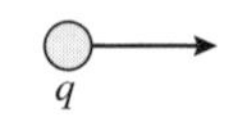

   a. If q is a positive charge, does it speed up or slow down as it approaches the stationary charge? Answer this question twice:

      i. Using the concept of force.

      ii. Using the concept of energy.

   b. Repeat part a for q as a negative charge.

4. The figure shows two capacitors, each with a 3 mm separation. A proton is released from rest in the center of each capacitor.

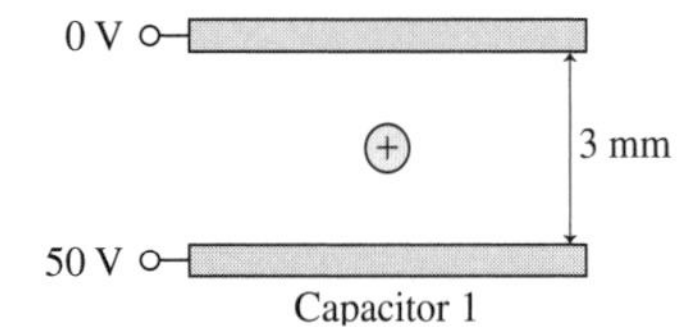

Capacitor 1

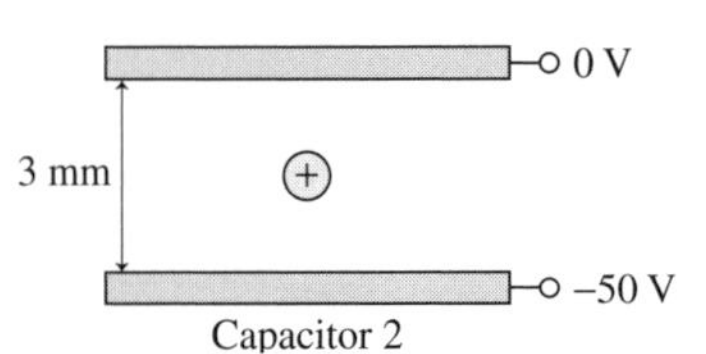

Capacitor 2

   a. Draw an arrow on each proton to show the direction it moves.

   b. Which proton reaches a capacitor plate first? Or are they simultaneous? Explain.

21.3 Calculating the Electric Potential

5. Rank in order, from largest to smallest, the electric potentials V_1 to V_5 at points 1 to 5.

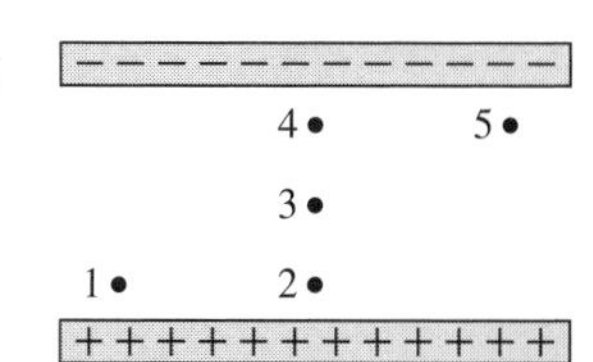

Order:

Explanation:

6. The figure shows the potential energy of a positively charged particle in a region of space.

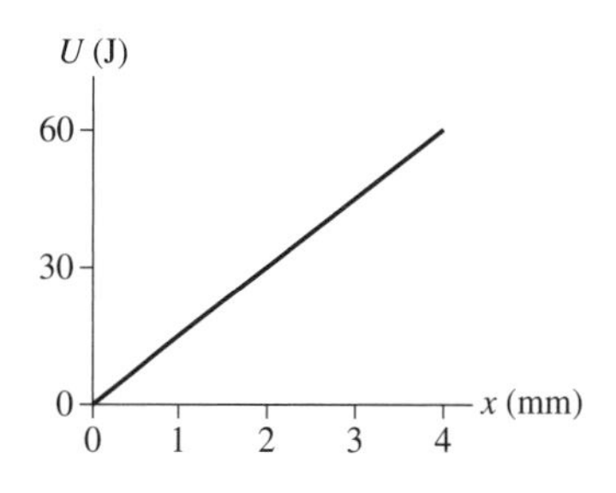

a. What arrangement of source charges is responsible for this potential energy? Draw the source above the axis below.

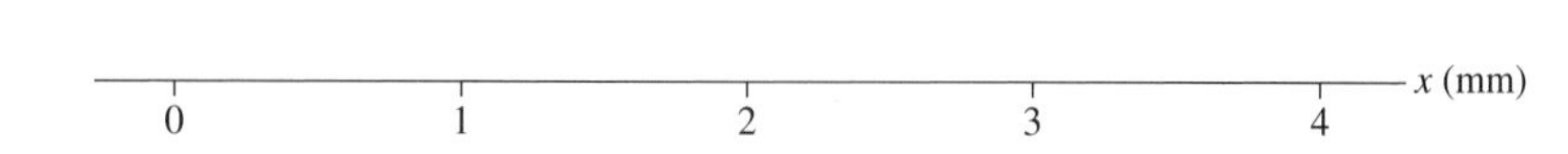

b. With what kinetic energy should the charged particle be launched from $x = 0$ mm to have a turning point at $x = 3$ mm? Explain.

c. How much kinetic energy does the charged particle of part b have as it passes $x = 2$ mm?

7. a. Charge $q_1 = 3$ nC is distance r from a positive point charge Q. Charge $q_2 = 1$ nC is distance $2r$ from Q. What is the ratio U_1/U_2 of their potential energies due to their interactions with Q?

b. Charge $q_1 = 3$ nC is distance d from the negative plate of a parallel-plate capacitor. Charge $q_2 = 1$ nC is distance $2d$ from the negative plate. What is the ratio U_1/U_2 of their potential energies?

8. The figure shows the potential energy of a proton ($q = +e$) and a lead nucleus ($q = +82e$). The horizontal scale is in units of *femtometers*, where 1 fm = 1 femtometer = 10^{-15} m.

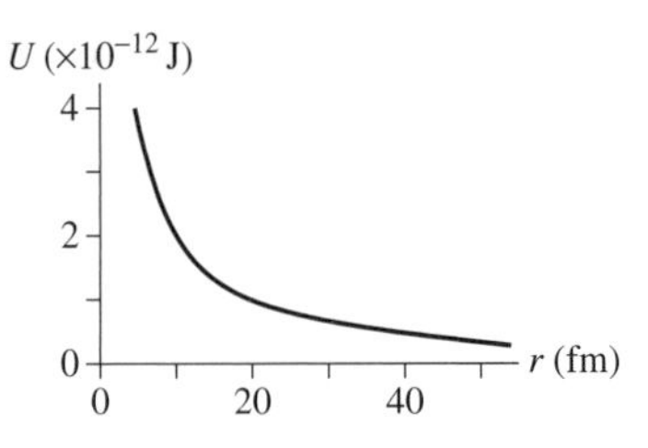

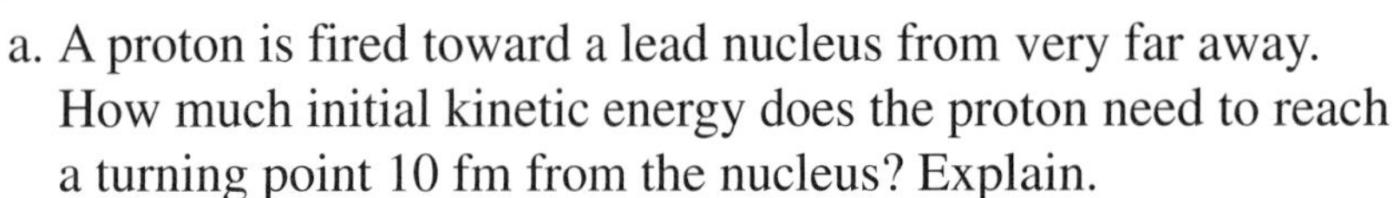

a. A proton is fired toward a lead nucleus from very far away. How much initial kinetic energy does the proton need to reach a turning point 10 fm from the nucleus? Explain.

b. How much kinetic energy does the proton of part a have when it is 20 fm from the nucleus and moving toward it, before the collision?

c. How much kinetic energy does the proton of part a have when it is 20 fm from the nucleus and moving away from it, after the collision?

9. Rank in order, from largest to smallest, the electric potentials V_1 to V_5 at points 1 to 5.

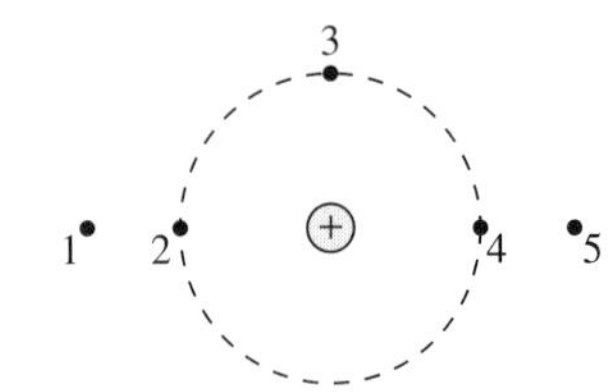

Order:

Explanation:

10. Each figure shows a contour map on the left and a set of graph axes on the right. Draw a graph of V versus x. Your graph should be a straight line or a smooth curve.

a.

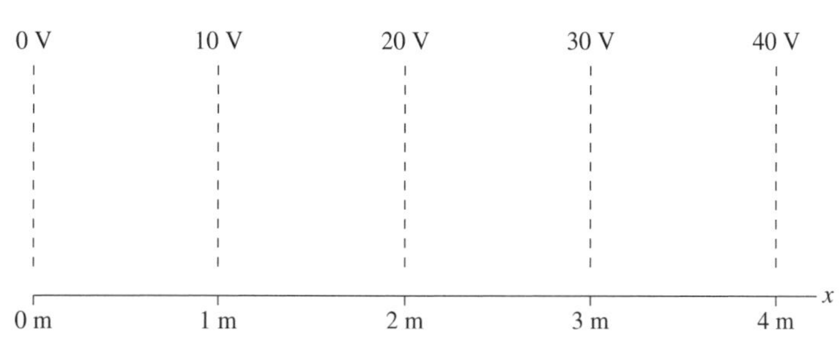

b.

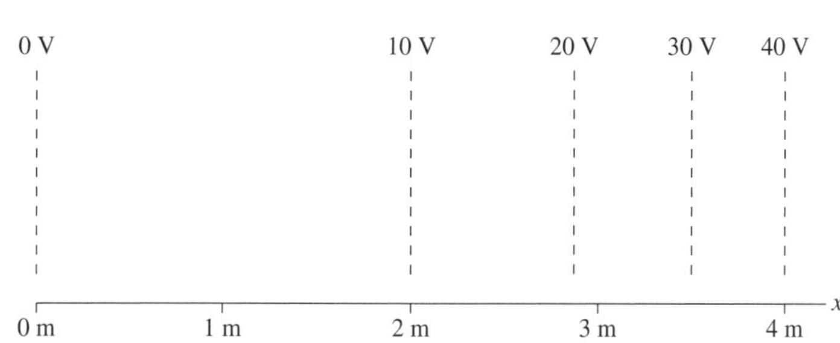

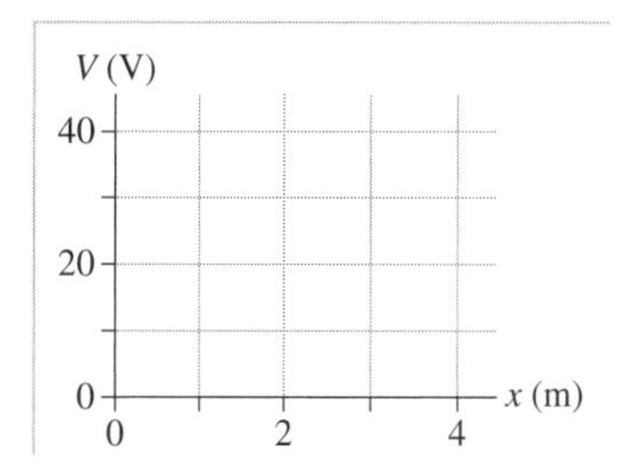

c.

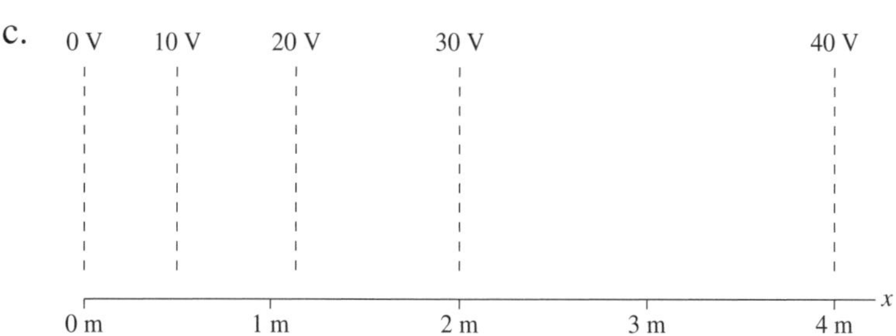

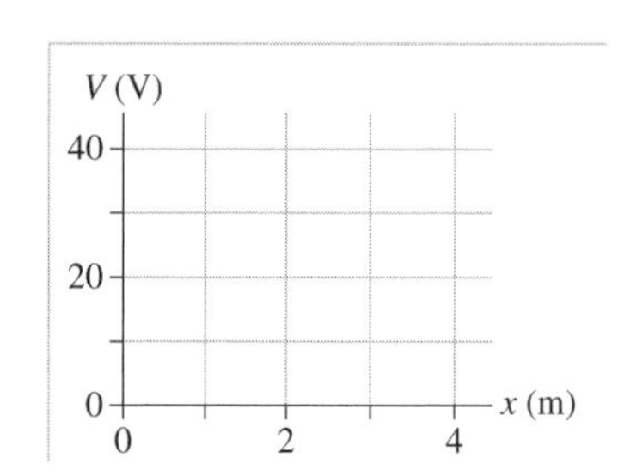

11. Each figure shows a V-versus-x graph on the left and an x-axis on the right. Assume that the potential varies with x but not with y. Draw a contour map of the electric potential. There should be a uniform potential difference between equipotential lines, and each equipotential line should be labeled.

a.

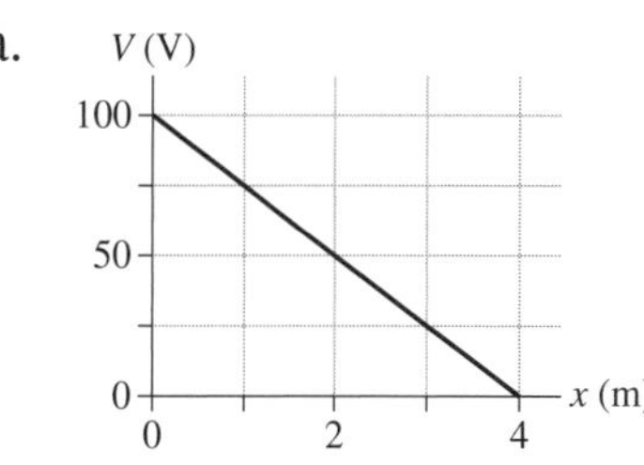

b.

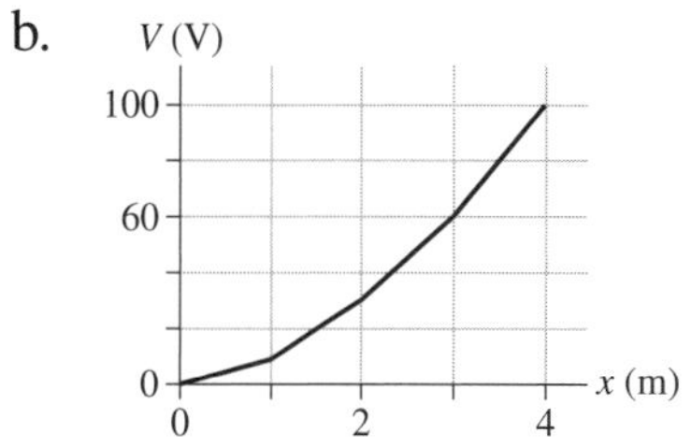

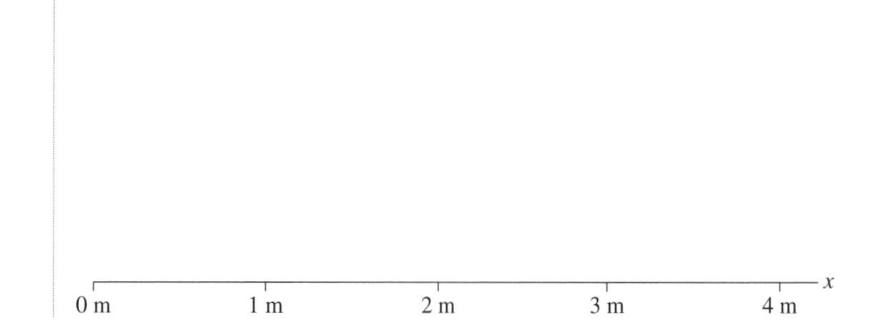

12. An inflatable metal balloon of radius R is charged to a potential of 1000 V. After all wires and batteries are disconnected, the balloon is inflated to a new radius $2R$.

    a. Does the potential of the balloon change as it is inflated? If so, by what factor? If not, why not?

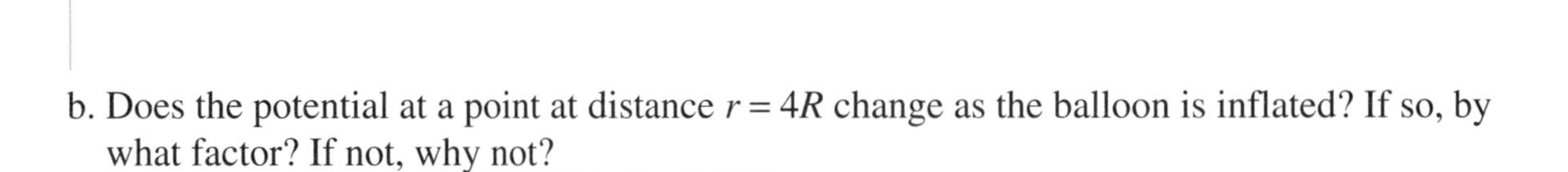

    b. Does the potential at a point at distance $r = 4R$ change as the balloon is inflated? If so, by what factor? If not, why not?

13. On the axes below, draw a graph of V versus x for the two point charges shown.

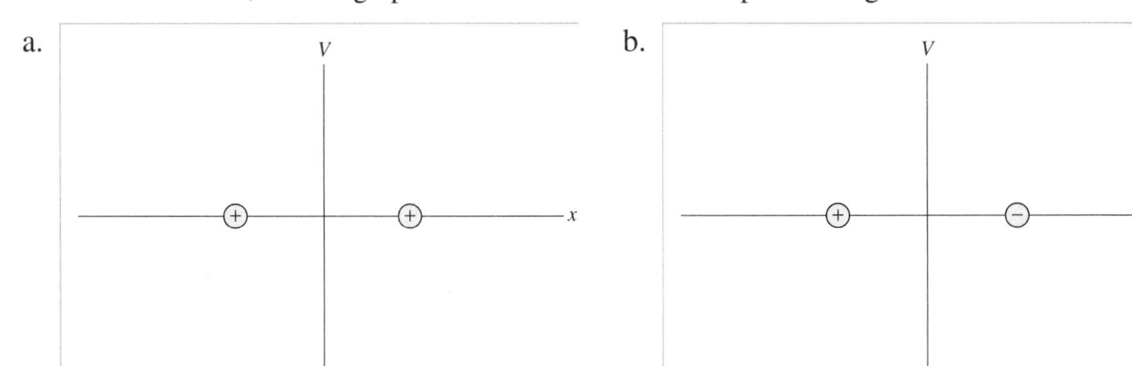

14. For each pair of charges below, are there any points (other than at infinity) at which the electric potential is zero? If so, identify them on the figure with a dot and a label. If not, why not?

a.

b.

21.4 Sources of Electric Potential

21.5 Connecting Potential and Field

15. The top graph on the right shows the electric potential as a function of x. On the axes below the graph, draw the graph of E_x versus x in this same region of space.

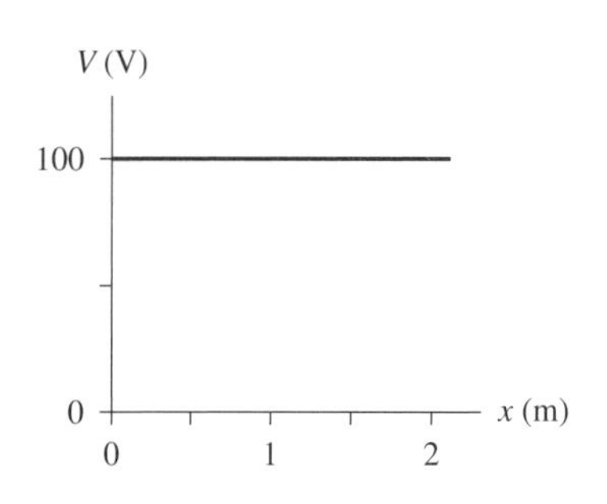

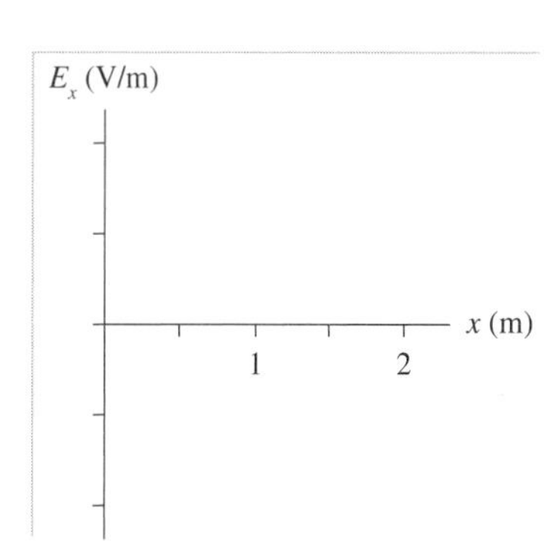

16. For each contour map:
    i. Estimate the electric fields $\vec{E}_a$ and $\vec{E}_b$ at points a and b. Don't forget that $\vec{E}$ is a vector. Show how you made your estimate.
    ii. Draw electric field vectors on top of the contour map.

a.

0 V 10 V 20 V 30 V 40 V

a b

0 m 1 m 2 m 3 m 4 m x

$\vec{E}_a =$ ____________

$\vec{E}_b =$ ____________

b.

0 V 10 V 20 V 30 V 40 V

a b

0 m 1 m 2 m 3 m 4 m x

$\vec{E}_a =$ ____________

$\vec{E}_b =$ ____________

17. Draw the electric field vectors at the dots on this contour map. The length of each vector should be proportional to the field strength at that point.

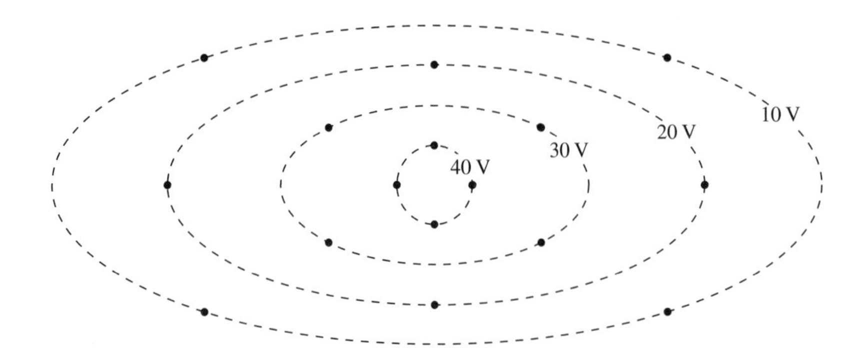

18. Draw the electric field vectors at the dots on this contour map. The length of each vector should be proportional to the field strength at that point.

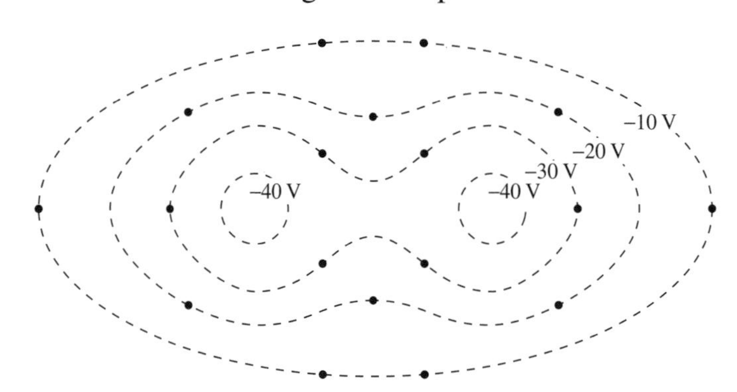

19. For each of the figures below, is this a physically possible potential map if there are no free charges in this region of space? If so, draw an electric field line diagram on top of the potential map. If not, why not?

a.

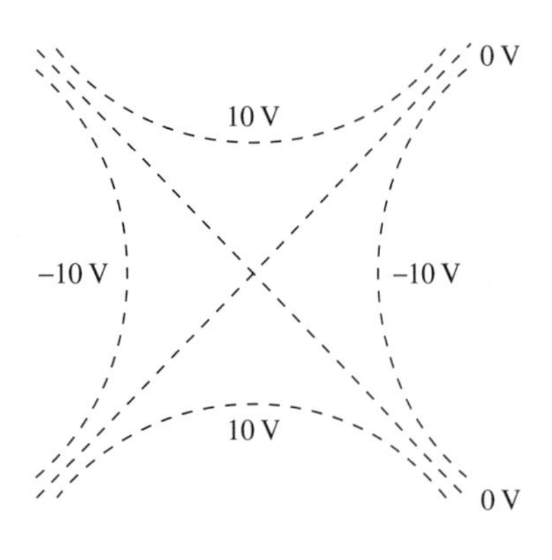

b.

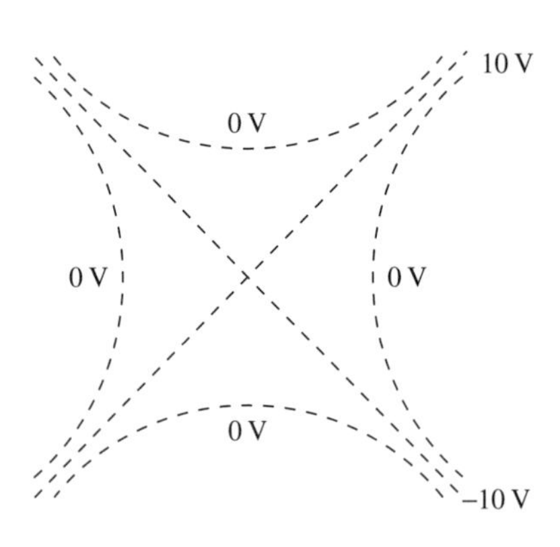

20. Two metal spheres are connected by a metal wire that has a switch in the middle. Initially the switch is open. Sphere 1, with the larger radius, is given a positive charge. Sphere 2, with the smaller radius, is neutral. Then the switch is closed. Afterward, sphere 1 has charge Q_1, is at potential V_1, and the electric field strength at its surface is E_1. The values for sphere 2 are Q_2, V_2, and E_2.

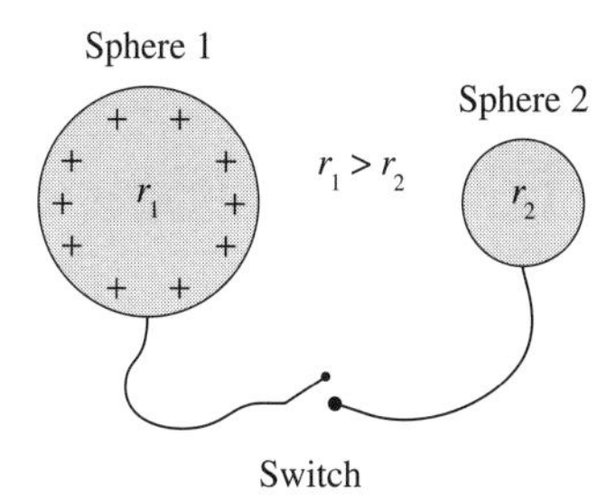

a. Is V_1 larger than, smaller than, or equal to V_2? Explain.

b. Is Q_1 larger than, smaller than, or equal to Q_2? Explain.

c. Is E_1 larger than, smaller than, or equal to E_2? Explain.

21. The figure shows a hollow metal shell. A negatively charged rod touches the top of the sphere, transferring charge to the shell. Then the rod is removed.

a. Show on the figure the equilibrium distribution of charge.

b. Draw the electric field diagram.

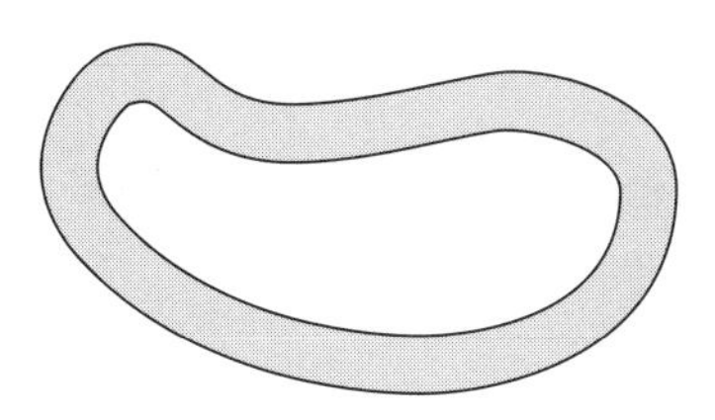

21.6 The Electrocardiogram

21.7 Capacitance and Capacitors

22. Rank in order, from largest to smallest, the potential differences $(\Delta V_C)_1$ to $(\Delta V_C)_4$ of these four capacitors.

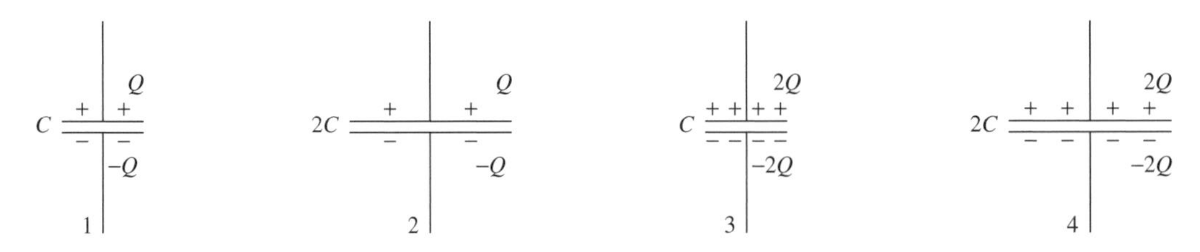

Order:

Explanation:

21.8 Dielectrics and Capacitors

21.9 Energy and Capacitors

23. A parallel plate capacitor is fully charged and disconnected from its charging source when a dielectric is inserted between the plates.
    a. Sketch the electric field lines between the plates to show the effect of the polarization of the dielectric.
    b. Is the electric field between the plates larger, smaller, or the same as it was before the dielectric was inserted? Explain.

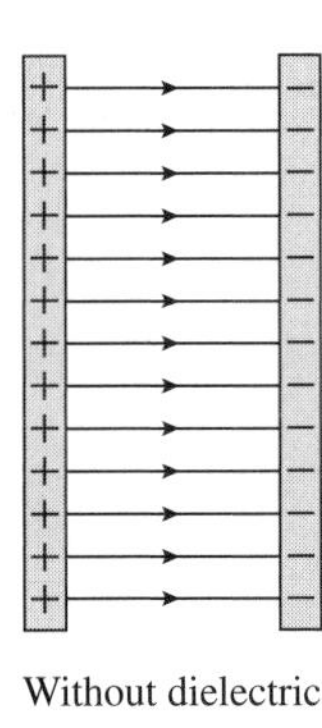

Without dielectric

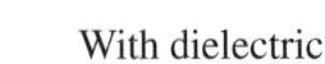

With dielectric

c. Is the potential difference between the plates larger, smaller, or the same as before the dielectric was inserted? Explain.

d. Is the energy stored by the capacitor larger, smaller, or the same as before the dielectric was inserted? Explain.

22 Current and Resistance

22.1 A Model of Current

22.2 Defining and Describing Current

1. a. Describe an experiment that provides evidence that current consists of charge flowing through a conductor. Use both pictures and words.

   b. One model of current is that it consists of the motion of discrete charged particles. Another model is that current is the flow of a continuous charged fluid. Does the experiment you described in part a provide evidence in favor of either one of these models? If so, describe how.

2. Figure A shows capacitor plates that have been charged to $+/-Q$. A very long conducting wire is then connected to the plates as shown in Figure B. Will the separated charges remain on the plates where they are attracted to each other by Coulomb attraction, or will the charges travel through the long wires to discharge the plates? Explain.

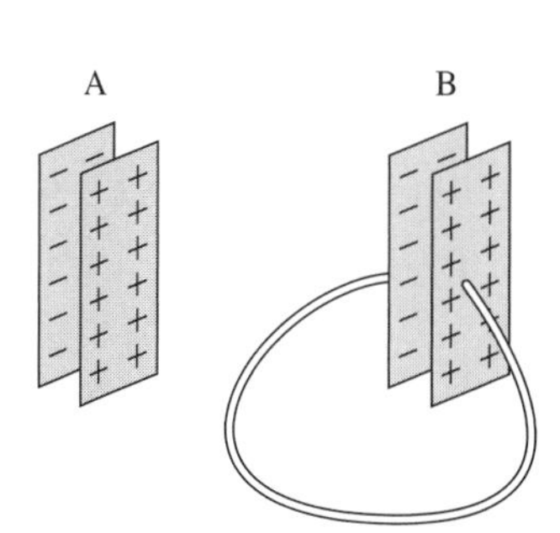

3. A discharging capacitor is used to light bulbs of significantly different types and at different locations as shown.

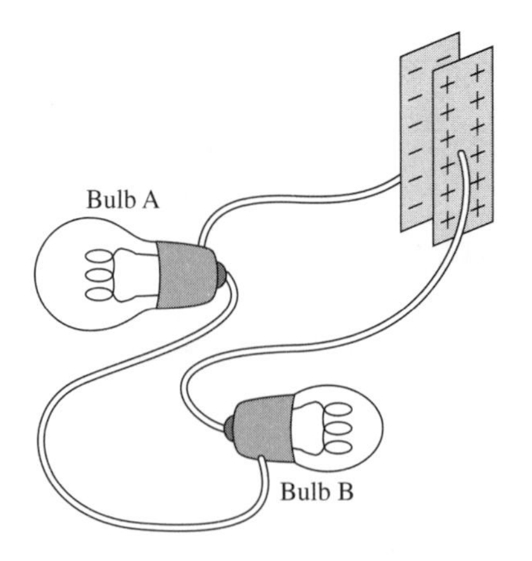

   a. Which of the following characteristics, if any, will cause the amount of current in Bulb A to be different than that through Bulb B? Explain.
      - Bulb A is rated at a higher wattage than Bulb B.
      - Bulb A is connected directly to the negative plate, while Bulb B is connected to the positive plate.
      - Bulb A is closer (connected through less wire) to the capacitor plates than Bulb B.

   b. Which of the following characteristics, if any, will cause Bulb A to light before Bulb B? Which will cause Bulb B to light before Bulb A? Explain.
      - Bulb A is rated at a higher wattage than Bulb B.
      - Bulb A is connected directly to the negative plate, while Bulb B is connected to the positive plate.
      - Bulb A is closer to the capacitor plates than Bulb B.

4. The figure shows a segment of a current-carrying metal wire.

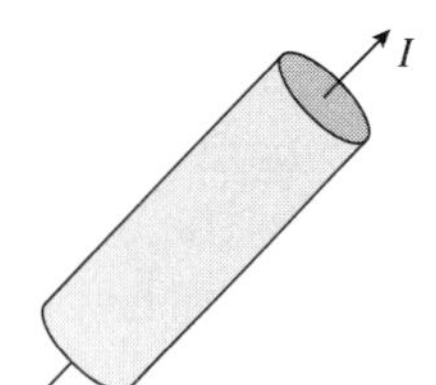

   a. Draw an arrow on the figure, using a **black** pen or pencil, to show the direction of motion of the charge carriers.
   b. Draw an arrow on the figure, using a **red** pen or pencil, to show the direction of the electric field.

22.3 Batteries and EMF

22.4 Connecting Potential and Current

5. A light bulb is connected to a battery with 1-mm-diameter wires. The bulb is glowing.
   a. Draw arrows at points 1, 2, and 3 to show the direction of the electric field at those points. (The points are *inside* the wire.)
   b. Rank in order, from largest to smallest, the field strengths E_1, E_2, and E_3.

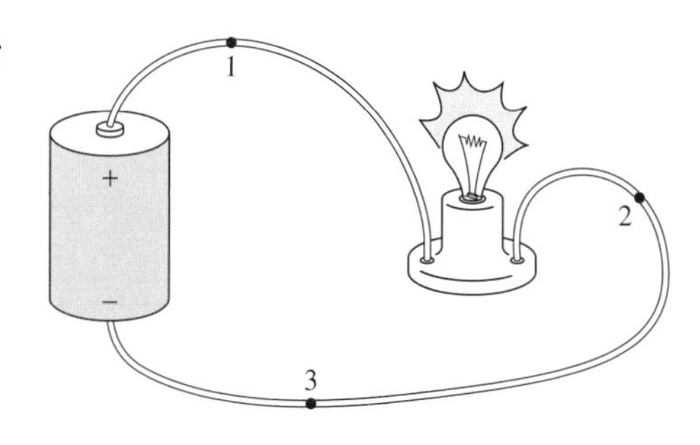

Order:

Explanation:

6. Wire 1 and Wire 2 are made of the same metal and are the same length. Wire 1 has twice the diameter and half the potential difference across its ends. What is the ratio of I_1/I_2?

7. If a metal is heated, does its resistivity increase, decrease, or stay the same? Explain.

22.5 Resistors and Ohm's Law

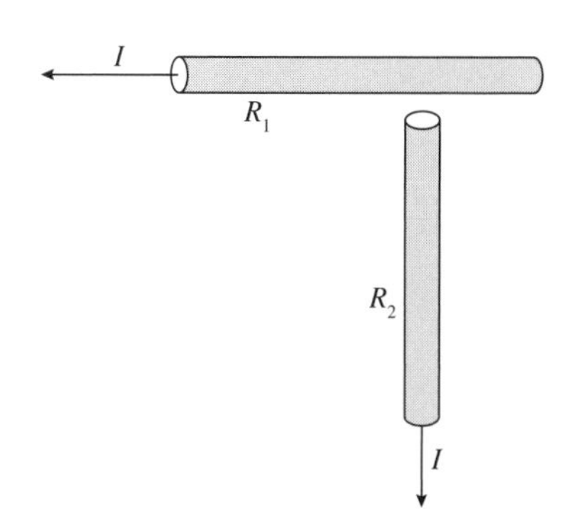

8. For resistors R_1 and R_2:
   a. Which end (left, right, top, or bottom) is more positive?

   R_1 ________ R_2 ________

   b. In which direction (such as left to right or top to bottom) does the potential decrease?

   R_1 ________

   R_2 ________

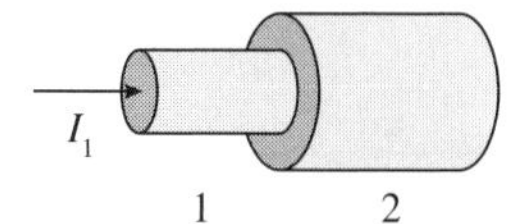

9. A wire consist of two segments of different diameters but made from the same metal. The current in segment 1 is I_1.
   a. Compare the values of the currents in the two segments. Is I_2 greater than, less than, or equal to I_1? Explain.

   b. Compare the strengths of the electric fields E_1 and E_2 in the two segments.

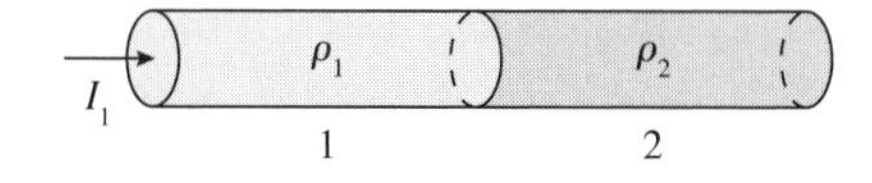

10. A wire consists of two equal-diameter segments. Their resistivities differ, with $\rho_1 > \rho_2$. The current in segment 1 is I_1.
    a. Compare the values of the currents in the two segments. Is I_2 greater than, less than, or equal to I_1? Explain.

    b. Compare the strengths of the electric fields E_1 and E_2 in the two segments.

11. A graph of current as a function of potential difference is given for a particular wire segment.
    a. What is the resistance of the wire?

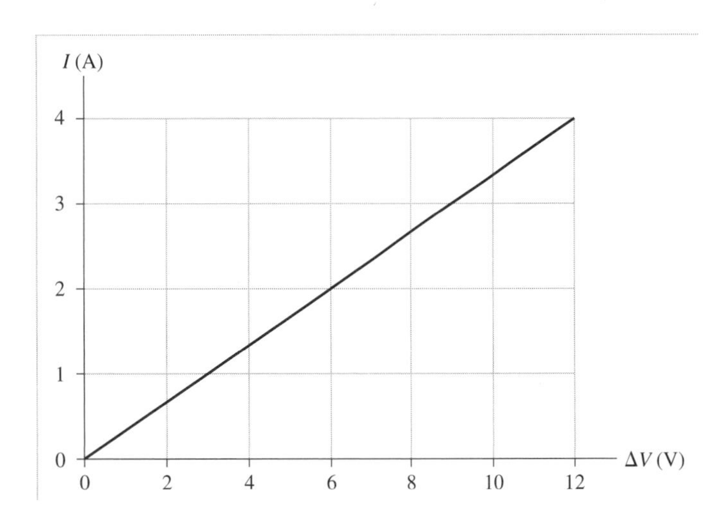

    b. Sketch and label on the same axes the graph of current vs. potential difference for a wire made of the same material, but twice as long as the wire in part a.
    c. Sketch and label on the same axes the graph of current vs. potential difference for a wire made of the same material, but with twice the cross-sectional area of the wire in part a.

12. A current vs. potential difference graph is shown for a material kept at a constant temperature. Sketch on the graph how the plot might change if the material's resistivity increases with temperature and the material's temperature increases with increasing current.

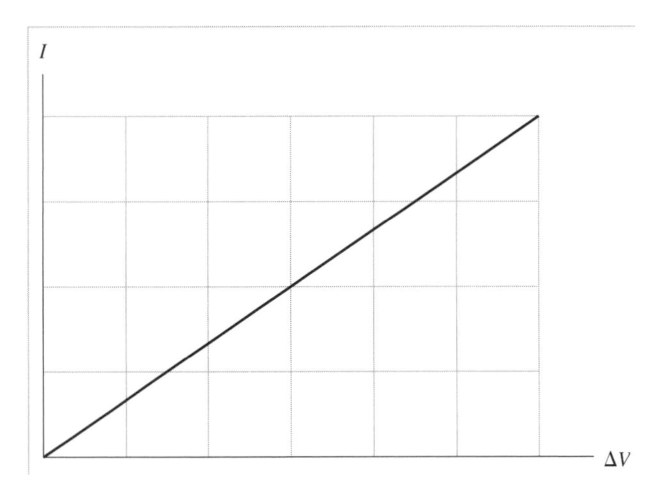

22.6 Energy and Power

13. a. Two conductors of equal lengths are connected in a line so that the same current I flows through them. The conductors are made of the same material but have different radii. Which of the two conductors dissipates the larger amount of power? Explain.

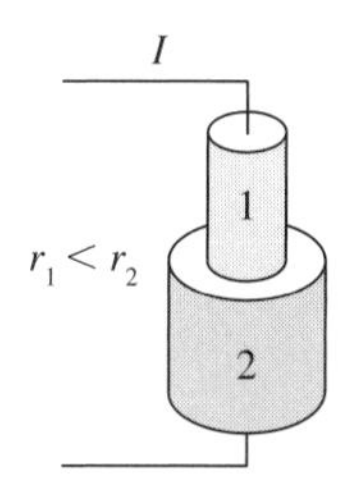

b. The same two conductors in part a are then each connected separately to identical batteries so that each has the same potential difference across it. Which of the two conductors now dissipates the larger amount of power? Explain.

14. Two resistors of equal lengths are connected to a battery by ideal wires. The resistors have the same radii but are made of different materials and have different resistivities ρ.

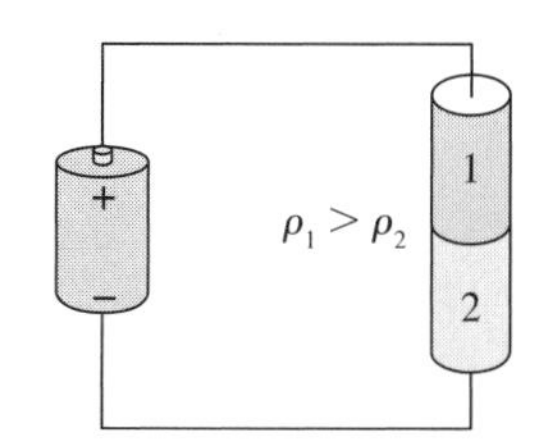

a. How does the current through each resistor compare?

b. Which of the two resistors dissipates the larger amount of power? Explain.

c. Is the voltage difference across the two resistors the same? Explain.

23 Circuits

23.1 Circuit Elements and Diagrams

Exercises 1–4: Redraw the circuits shown using standard circuit symbols with only right angle connections.

1.

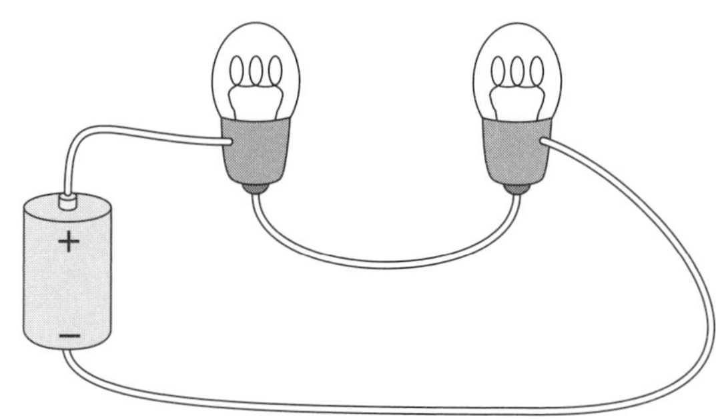

2.

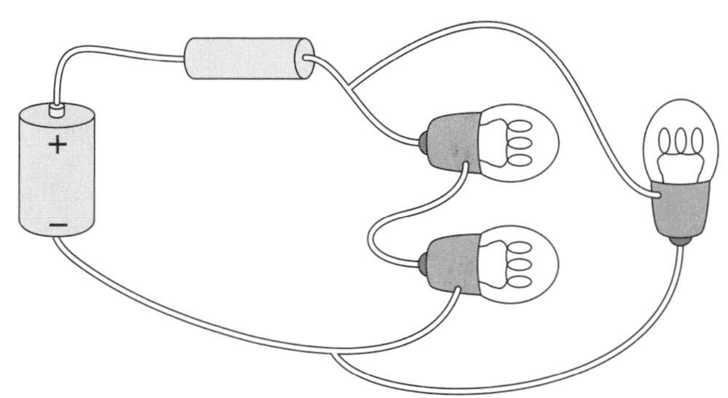

3.

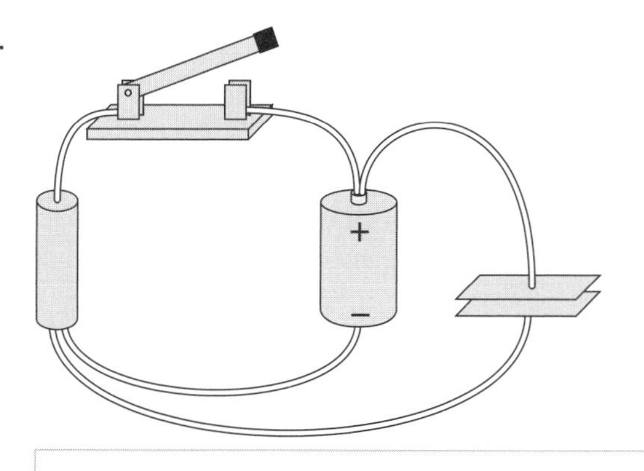

4.

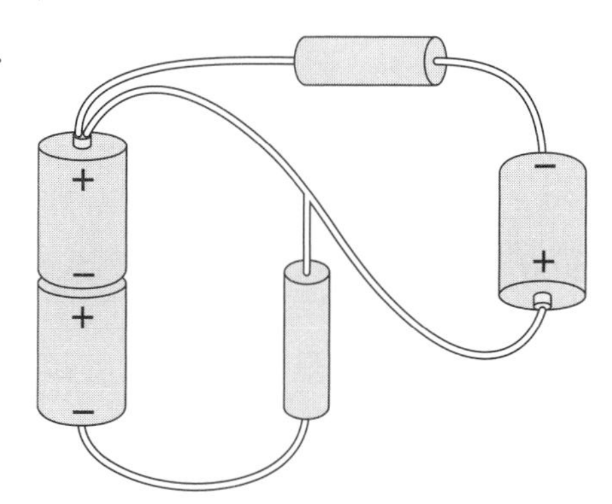

23.2 Kirchhoff's Laws

5. a. Redraw the circuit shown as a standard circuit diagram.

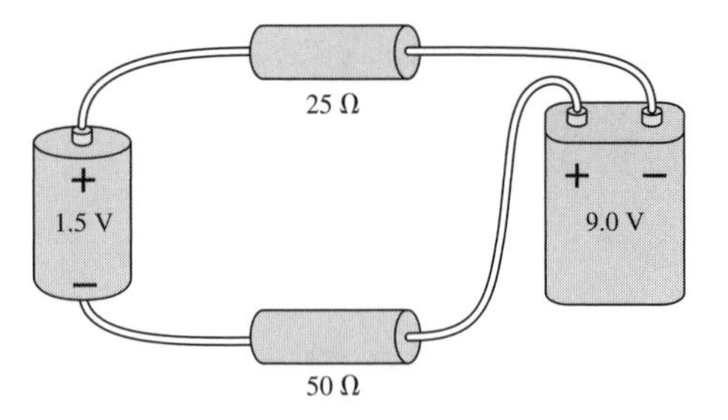

b. Assign a direction for the current and label the current arrow I on your sketch.
c. Apply Kirchoff's laws to determine the current through the resistors.

6. Draw a circuit for which the Kirchhoff loop law equation is

$$6\,\text{V} - I \cdot 2\,\Omega + 3\,\text{V} - I \cdot 4\,\Omega = 0$$

Assume that the analysis is done in a clockwise direction.

23.3 Series and Parallel Circuits

7. The figure shows five combinations of identical resistors. Rank in order, from largest to smallest, the equivalent resistances $(R_{eq})_1$ to $(R_{eq})_5$.

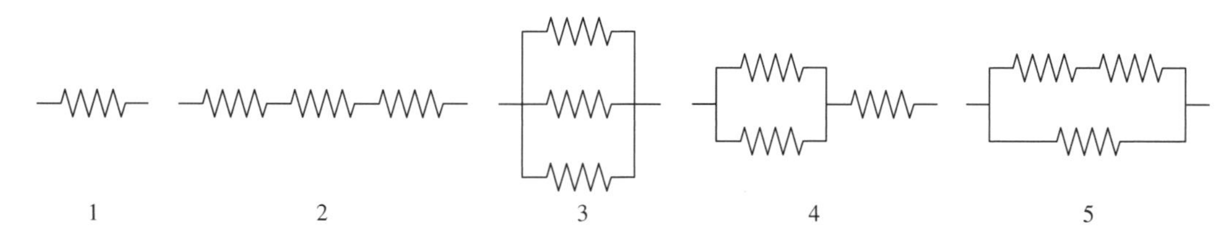

Order:

Explanation:

8. A 60 W light bulb and a 100 W light bulb are placed one after the other in a circuit. The battery's emf is large enough that both bulbs are glowing. Which one glows more brightly? Explain.

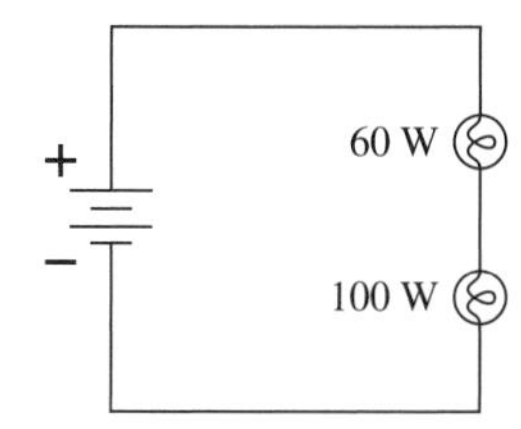

9. In your own words, without using a mathematical formula, state why the equivalent resistance of any number of resistors in series is always greater than any of the individual resistances and why the equivalent resistance of any number of resistances in parallel is always less than any of the individual resistances.

10. Are the three resistors shown wired in series, parallel, or a combination of series and parallel? To test your conclusion, trace all connections to the positive terminal of the battery in red. (These are all at the same potential.) Now trace all connections to the negative side of the battery in blue. (These are also at a common potential.) What do your tracings imply about how these three resistors are wired?

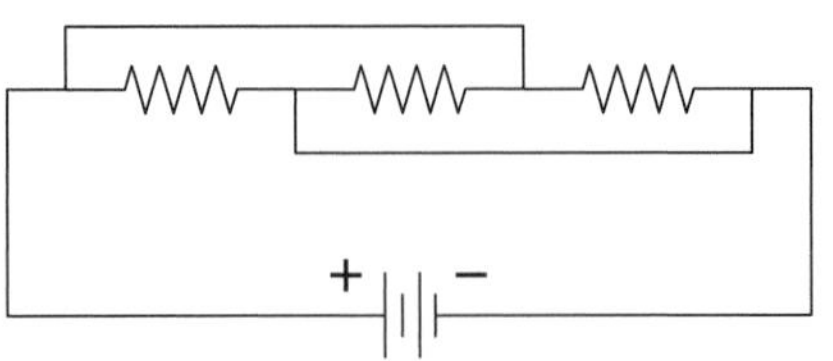

23.4 Measuring Voltage and Current

11. The diagram shows an incorrect way to attach an ammeter to a circuit. Why is this method not only wrong, but also potentially dangerous? What would the ammeter read?

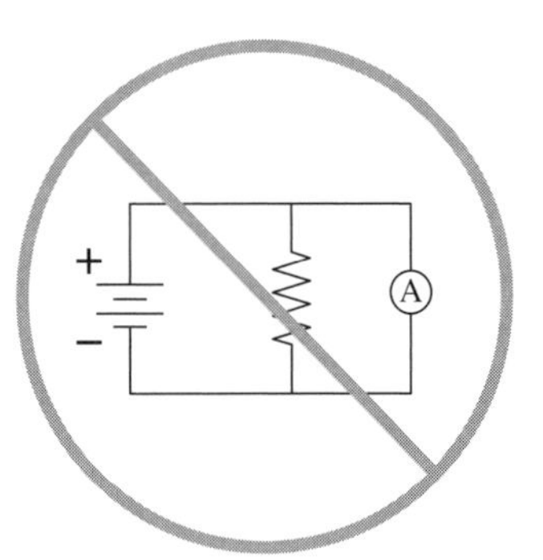

12. The diagram shows an incorrect way to attach a voltmeter to a circuit if the intent is to measure the potential difference across the 2 Ω resistor. What potential difference would the voltmeter read in this instance? How does that reading compare to the voltmeter reading expected from placing the voltmeter properly in parallel with the 2 Ω resistor?

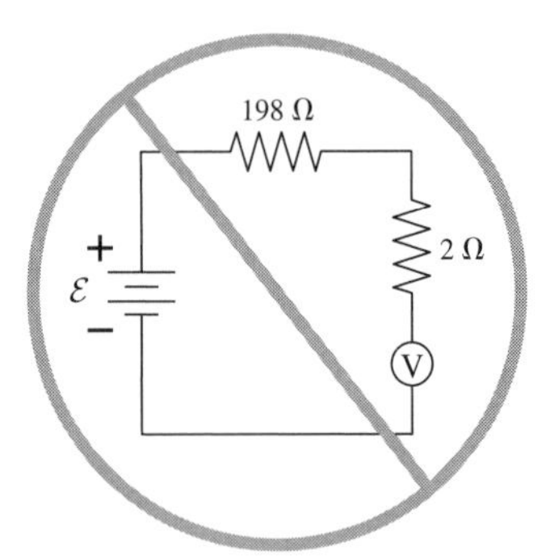

23.5 More Complex Circuits

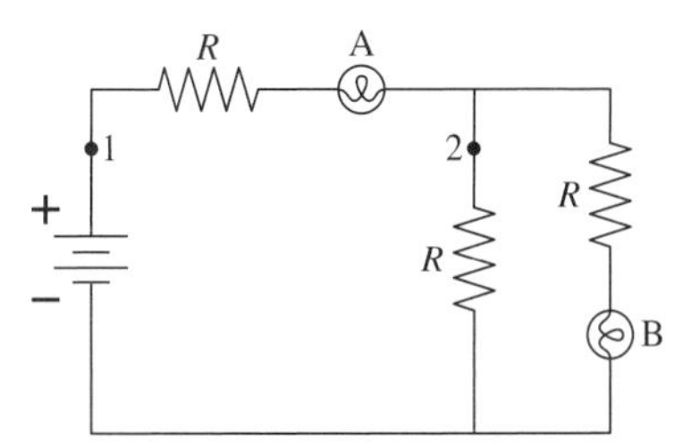

13. Bulbs A and B are identical. Initially both are glowing.
   a. Bulb A is removed from its socket. What happens to bulb B? Does it get brighter, stay the same, get dimmer, or go out? Explain.

   b. Bulb A is replaced. Bulb B is then removed from its socket. What happens to bulb A? Does it get brighter, stay the same, get dimmer, or go out? Explain.

   c. The circuit is restored to its initial condition. A wire is then connected between points 1 and 2. What happens to the brightness of each bulb?

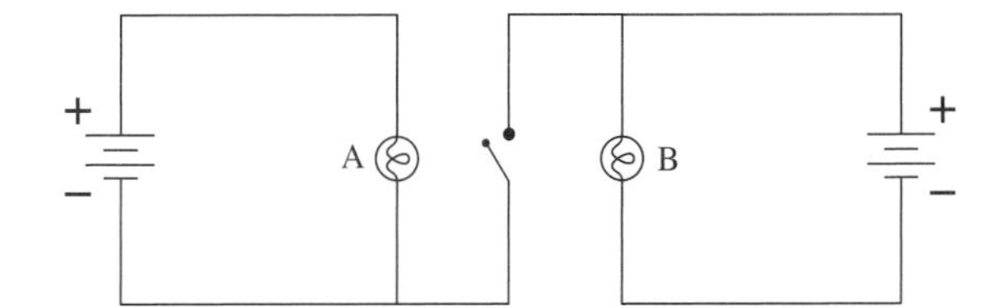

14. Bulbs A and B are identical and initially both are glowing. Then the switch is closed. What happens to each bulb? Does its brightness increase, stay the same, decrease, or go out? Explain.

23.6 Household Electricity

15. Bulbs A and B are identical and initially both are glowing. Then the switch is closed. What happens to each bulb? Does its brightness increase, stay the same, decrease, or go out? Explain.

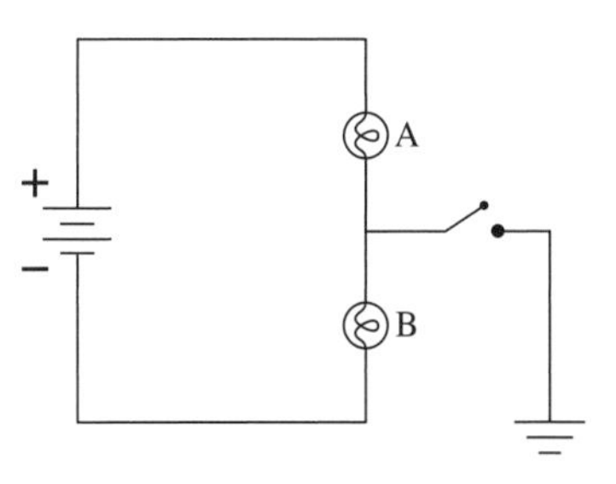

16. What is the difference, or is there a difference, between circuits 1 and 2? Explain.

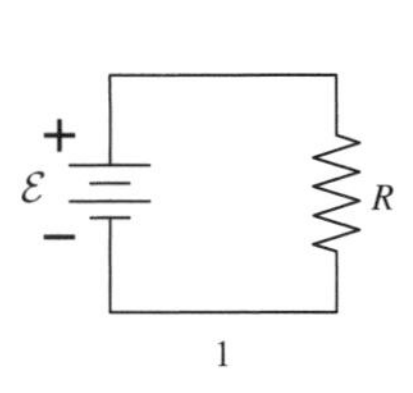

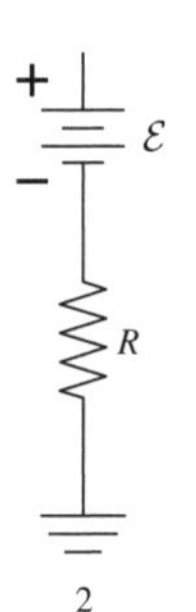

23.7 Capacitors in Parallel and Series

17. Each capacitor in the circuits below has capacitance C. What is the equivalent capacitance of the group of capacitors?

a.

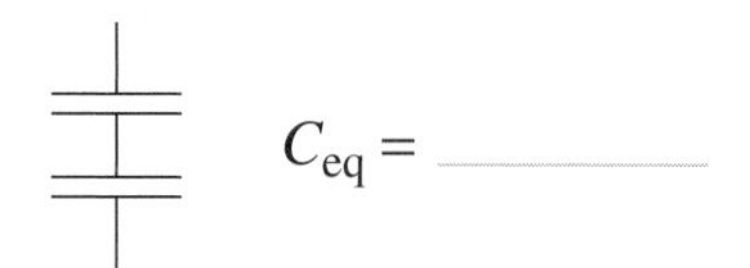

b.

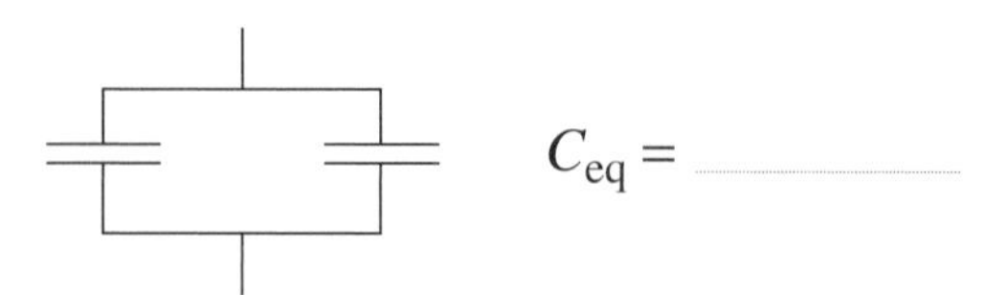

c.

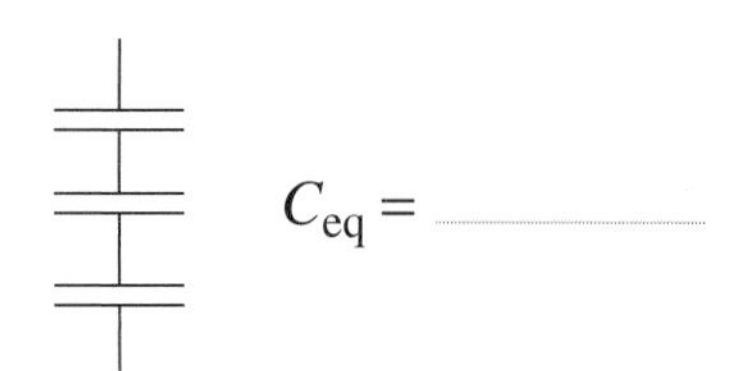

d.

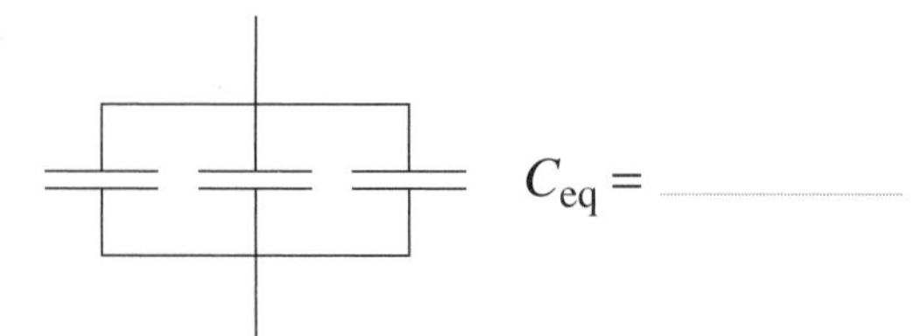

e.

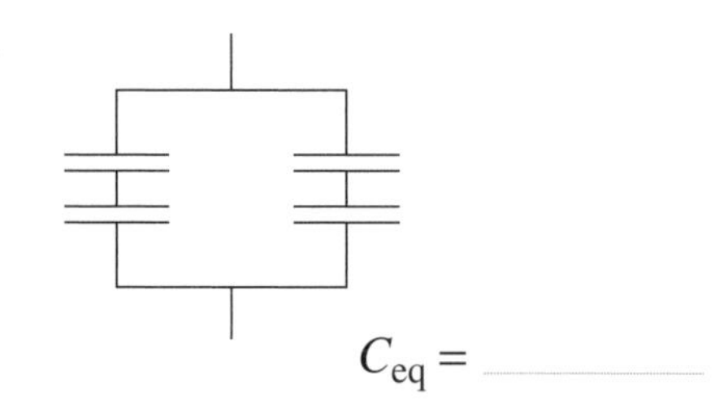

f.

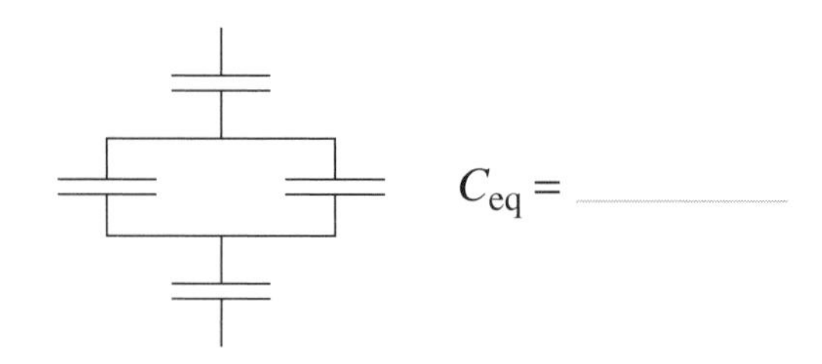

18. In your own words, without using a mathematical formula, state why the equivalent capacitance of any number of capacitors in series is always less than any of the individual capacitances and why the equivalent capacitance of any number of capacitors in parallel is always greater than any of the individual capacitances.

23.8 *RC* Circuits

19. The capacitors in each circuit are discharged when the switch closes at $t = 0$ s. Rank in order, from largest to smallest, the time constants τ_1 to τ_5 with which each circuit will discharge.

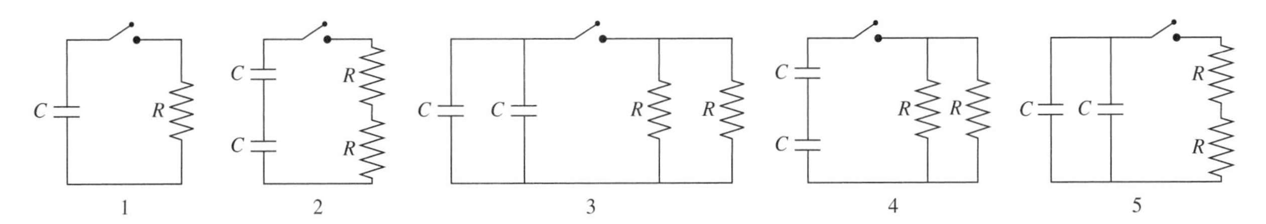

Order:

Explanation:

20. The charge on the capacitor is zero when the switch closes at $t = 0$ s.

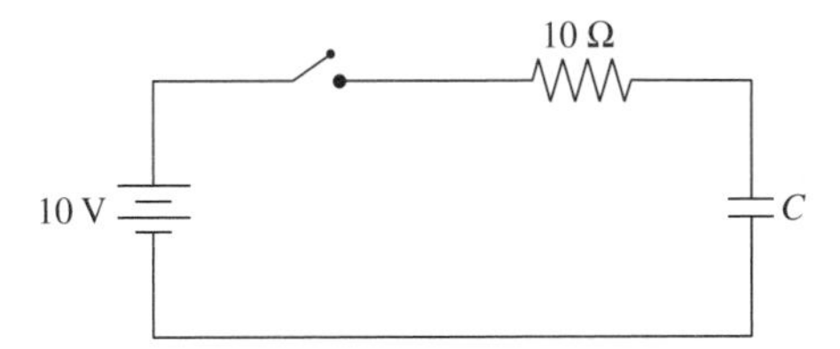

a. What will be the current in the circuit after the switch has been closed for a long time? Explain.

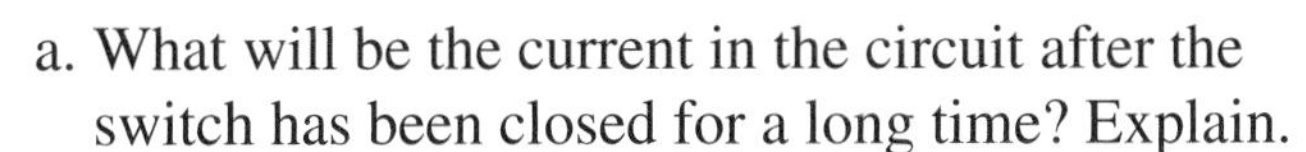

b. Immediately after the switch closes, before the capacitor has had time to charge, the potential difference across the capacitor is zero. What must be the potential difference across the resistor in order to satisfy Kirchhoff's loop law? Explain.

c. Based on your answer to part b, what is the current in the circuit immediately after the switch closes?

d. Sketch a graph of current versus time, starting from just before $t = 0$ s and continuing until the switch has been closed a long time. There are no numerical values for the horizontal axis, so you should think about the *shape* of the graph.

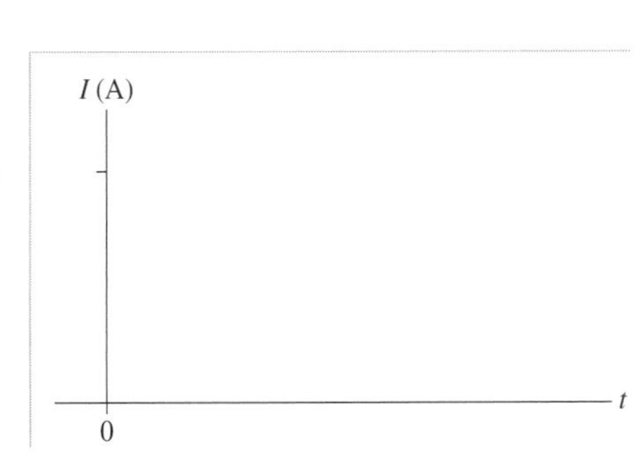

21. Graphs of the potential difference across the capacitor and resistor versus time are shown below for the *RC* circuit at right. Insert the appropriate times into the horizontal axes on the graphs and complete the third graph showing the sum of the voltages across capacitor and resistor.

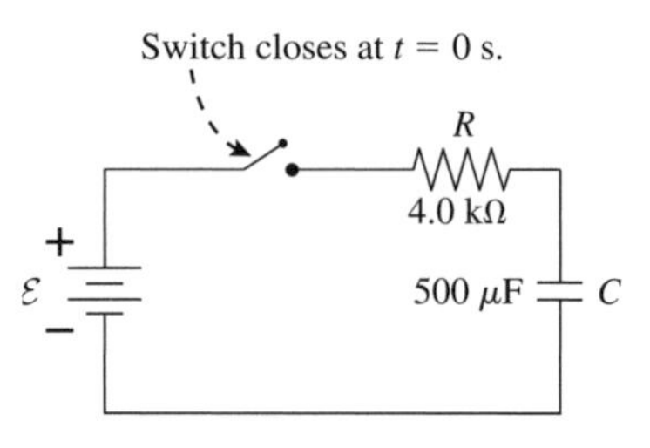

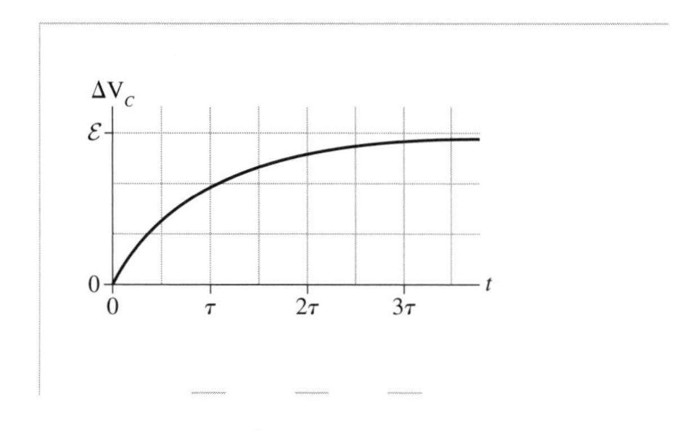

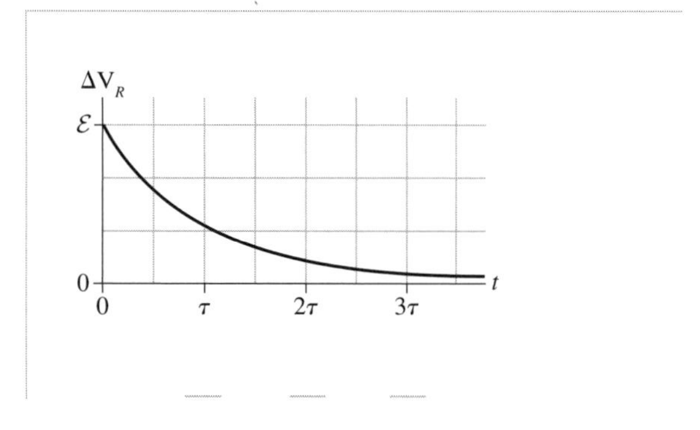

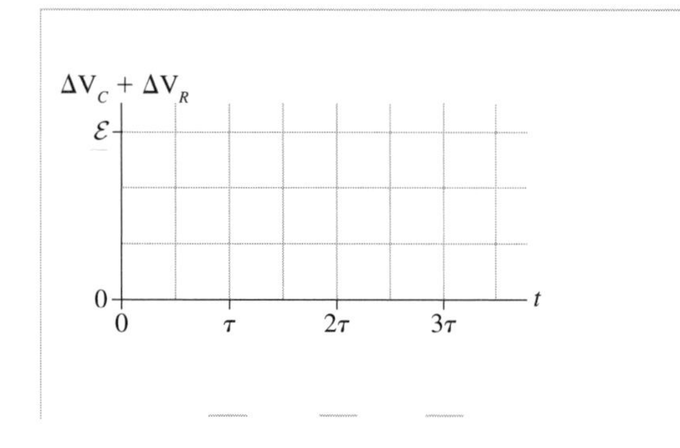

23.9 Electricity in the Nervous System

No exercises for this section.

24 Magnetic Fields and Forces

24.1 Magnetism

1. The compass needle below is free to rotate in the plane of the page. Either a bar magnet or a charged rod is brought toward the *center* of the compass. Does the compass rotate? If so, in which direction? If not, why not?

a.

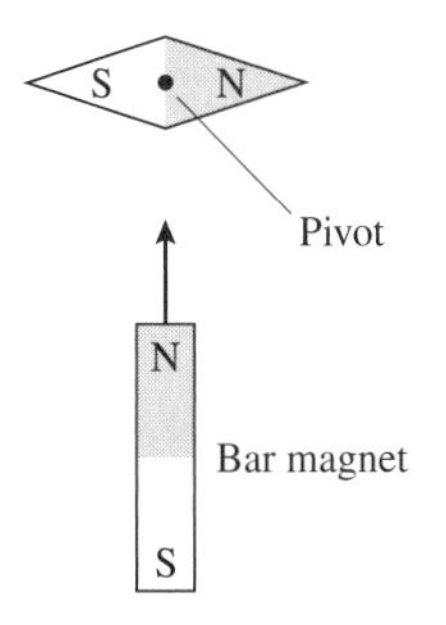

b.

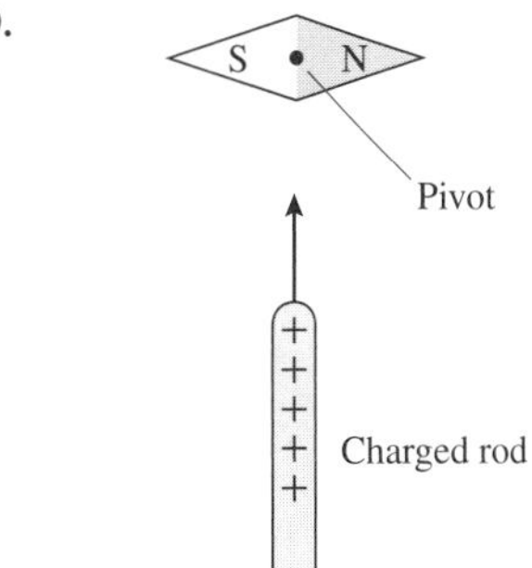

2. You have two electrically neutral metal cylinders that exert strong attractive forces on each other. You have no other metal objects. Can you determine if *both* of the cylinders are magnets, or if one is a magnet and the other just a piece of iron? If so, how? If not, why not?

3. Can you think of any kind of object that is repelled by *both* ends of a bar magnet? If so, what? If not, what prevents this from happening?

24.2 The Magnetic Field

4. A compass is placed at 12 different positions and its orientation is recorded. Use this information to draw the magnetic *field lines* in this region of space. Draw the field lines on the figure.

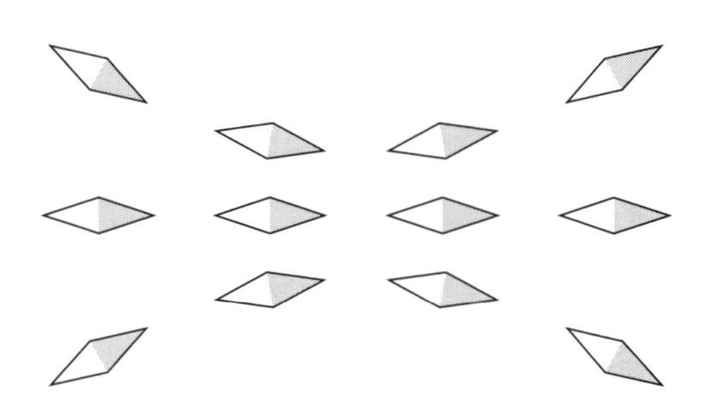

24.3 Electric Currents Also Create Magnetic Fields

5. A neutral copper rod, a charged insulator rod, and a bar magnet are arranged around a current-carrying wire as shown. For each, will it stay where it is? Move toward or away from the wire? Rotate clockwise or counterclockwise? Explain.

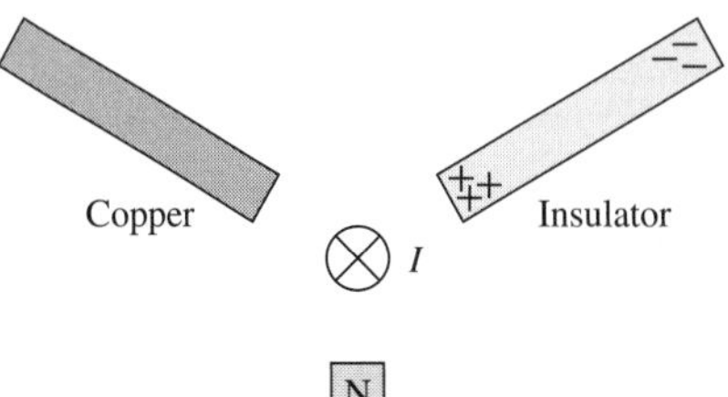

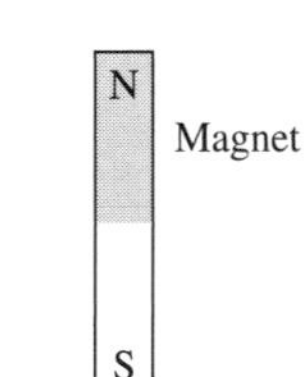

a. Neutral copper rod:

b. Insulating rod:

c. Bar magnet:

6. For each of the current-carrying wires shown, draw a compass needle in its equilibrium position at the positions of the dots. Label the poles of the compass needle.

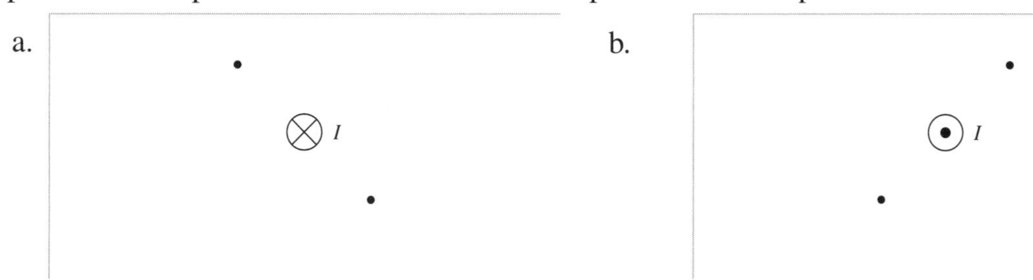

7. The figure shows a wire directed into the page and a nearby compass needle. Is the wire's current going into the page or coming out of the page? Explain.

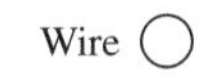

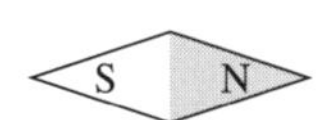

8. Each figure below shows a current-carrying wire. Draw the magnetic field diagram:

a.

The wire is perpendicular to the page. Draw magnetic field *lines*, then show the magnetic field *vectors* at a few points around the wire.

b.

The wire is in the plane of the page. Show the magnetic field above and below the wire.

9. This current-carrying wire is in the plane of the page. Draw the magnetic field on both sides of the wire.

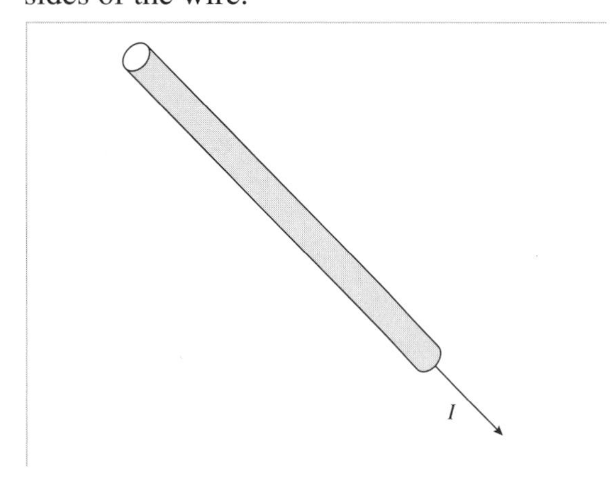

10. Use an arrow to show the current direction in this wire.

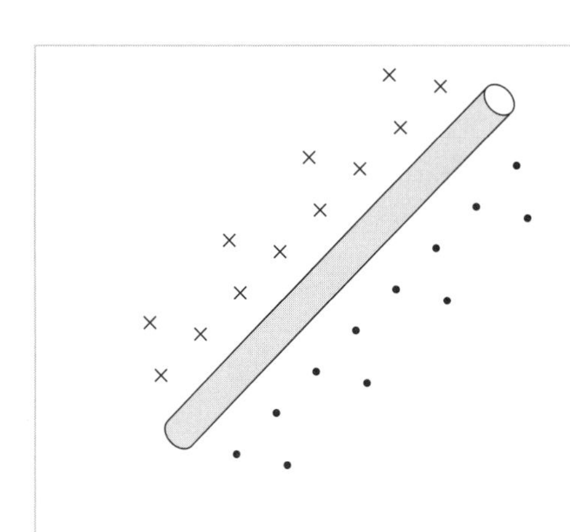

11. Each figure below shows two long straight wires carrying equal currents into and out of the page. At each of the dots, use a **black** pen or pencil to show and label the magnetic fields $\vec{B}_1$ and $\vec{B}_2$ of each wire. Then use a **red** pen or pencil to show the net magnetic field.

a.

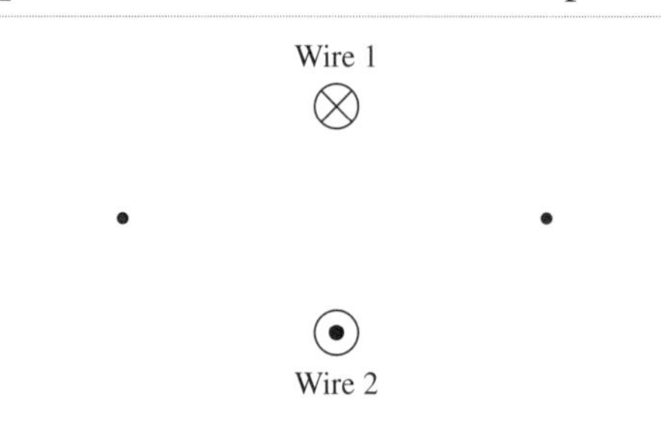

b.

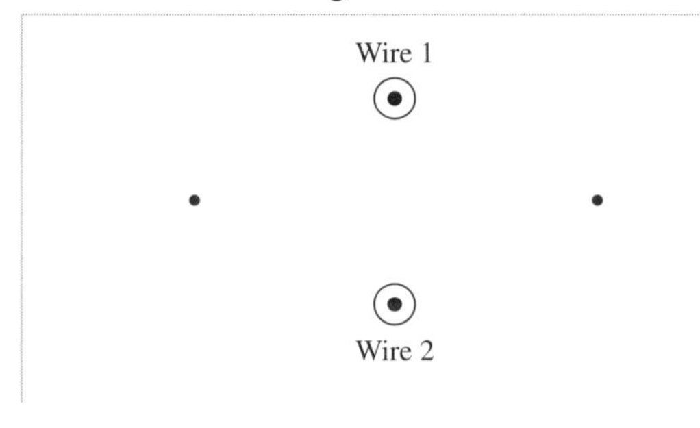

24.4 Calculating the Magnetic Field Due to a Current

12. A long straight wire, perpendicular to the page, passes through a uniform magnetic field. The *net* magnetic field at point 3 is zero.
    a. On the figure, show the direction of the current in the wire.
    b. Points 1 and 2 are the same distance from the wire as point 3, and point 4 is twice as distant. Construct vector diagrams at points 1, 2, and 4 to determine the net magnetic field at each point.

Uniform B-field

1

2

Wire

3

4

13. The figure shows the magnetic field seen when facing a current loop in the plane of the page. On the figure, show the direction of the current in the loop.

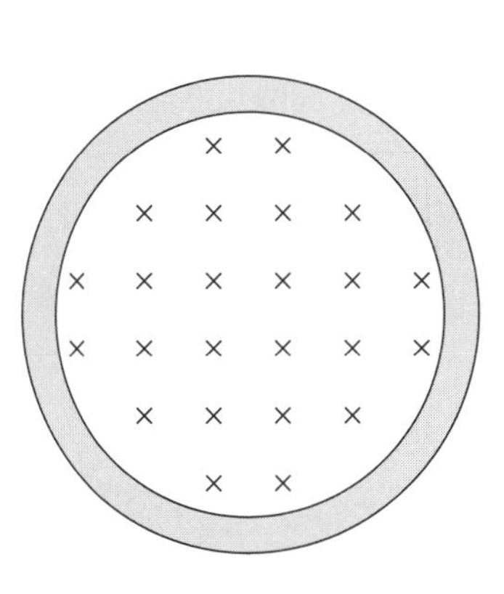

14. A solenoid with one layer of turns produces the magnetic field strength you need for an experiment when the current in the coil is 3 A. Unfortunately, this amount of current overheats the coil. You've determined that a current of 1 A would be more appropriate. How many additional layers of turns must you add to the solenoid to maintain the magnetic field strength?

15. Rank in order, from largest to smallest, the magnetic fields B_1 to B_3 produced by these three solenoids.

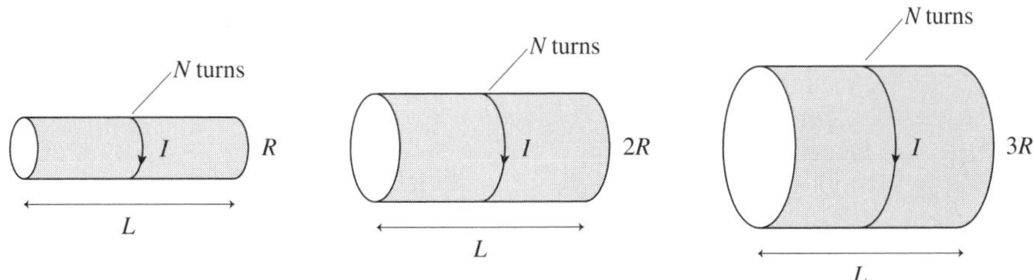

Order:

Explanation:

24.5 Magnetic Fields Exert Forces on Moving Charges

16. For each of the following, draw the magnetic force vector on the charge or, if appropriate, write "$\vec{F}$ into page," "$\vec{F}$ out of page," or "$\vec{F} = \vec{0}$."

a.
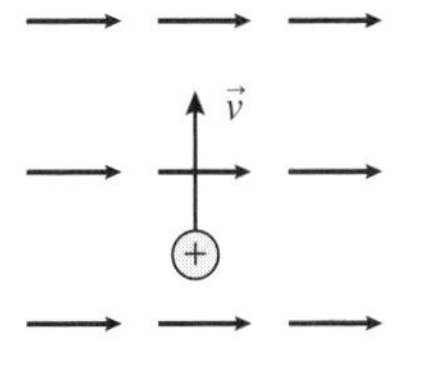

b.
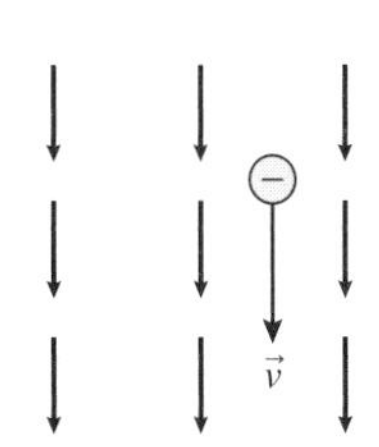

c.
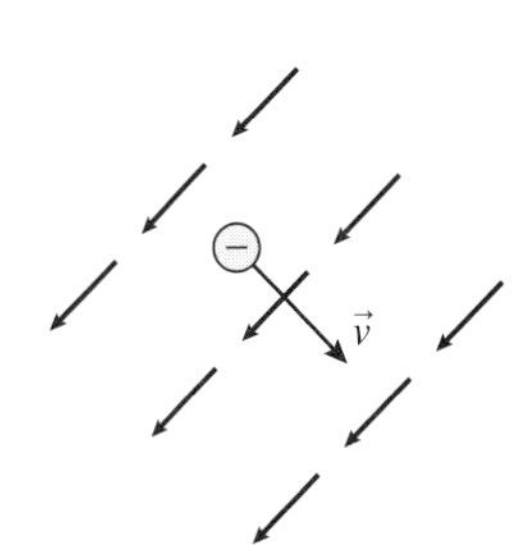

d.
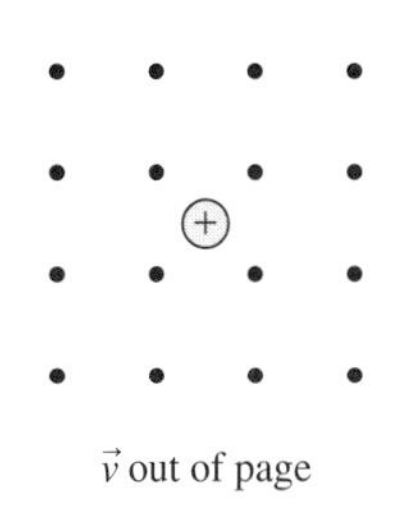

e.
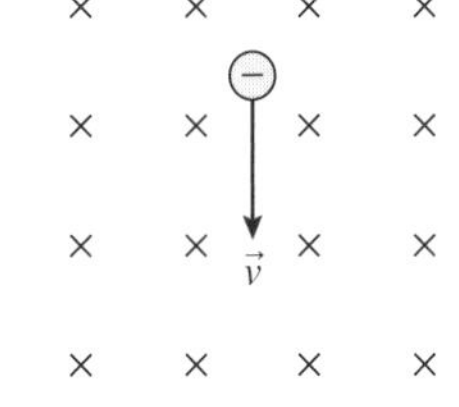

f.
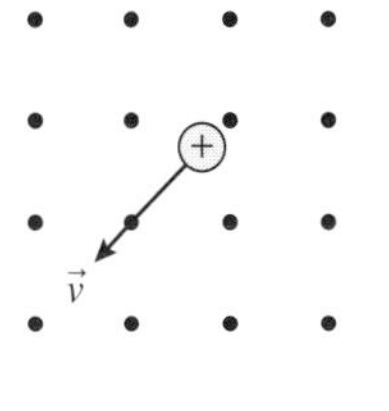

17. For each of the following, determine the sign of the charge (+ or −).

a.
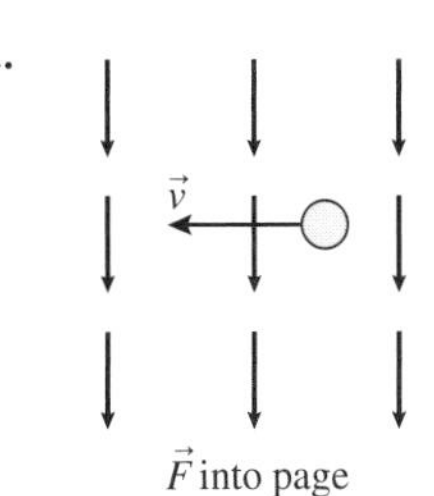

b.
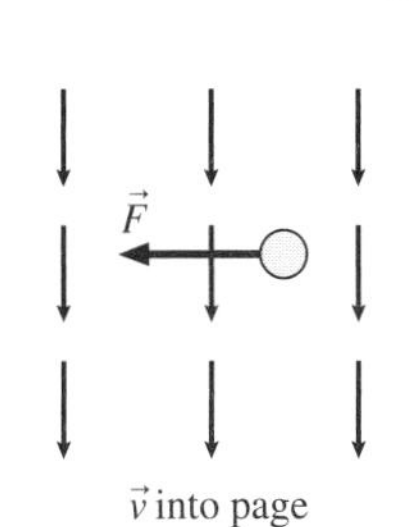

c.
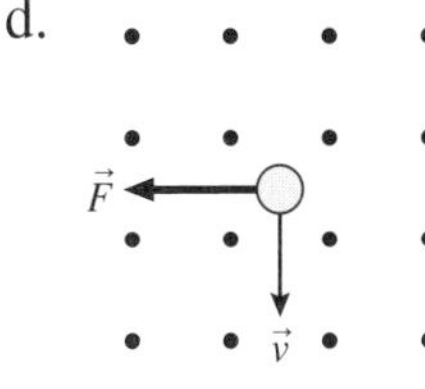

d.

$q =$ ______ $q =$ ______ $q =$ ______ $q =$ ______

18. The magnetic field is constant magnitude inside the dotted lines and zero outside. Sketch and label the trajectory of the charge for
   a. A weak field.
   b. A strong field.

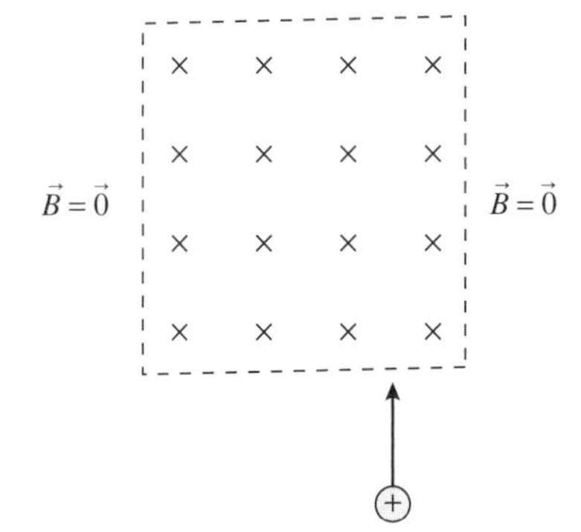

19. A positive ion, initially traveling into the page, is shot through the gap in a magnet. Is the ion deflected up, down, left, or right? Explain.

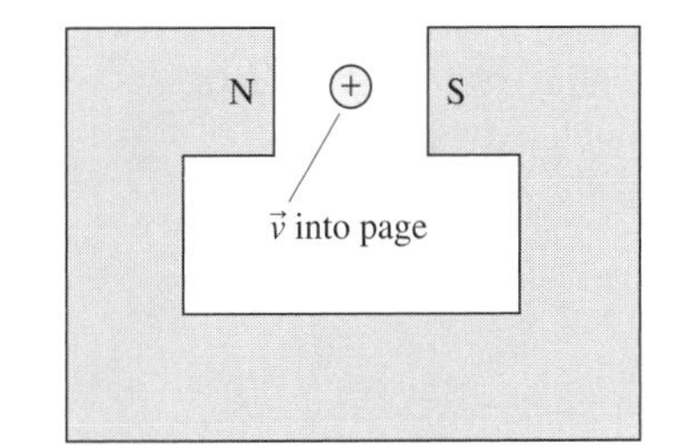

20. A positive ion is shot between the plates of a parallel-plate capacitor.

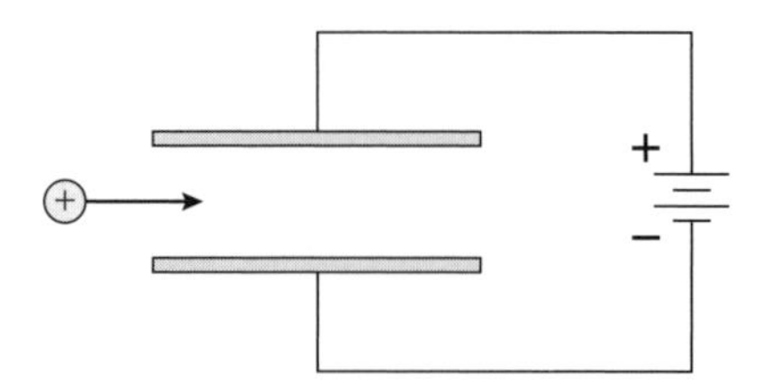

   a. In what direction is the electric force on the ion?

   b. Could a magnetic field exert a magnetic force on the ion that is opposite in direction to the electric force? If so, show the magnetic field on the figure.

21. In a high-energy physics experiment, a neutral particle enters a bubble chamber in which a magnetic field points into the page. The neutral particle undergoes a collision inside the bubble chamber, creating two charged particles. The subsequent trajectories of the charged particles are shown.

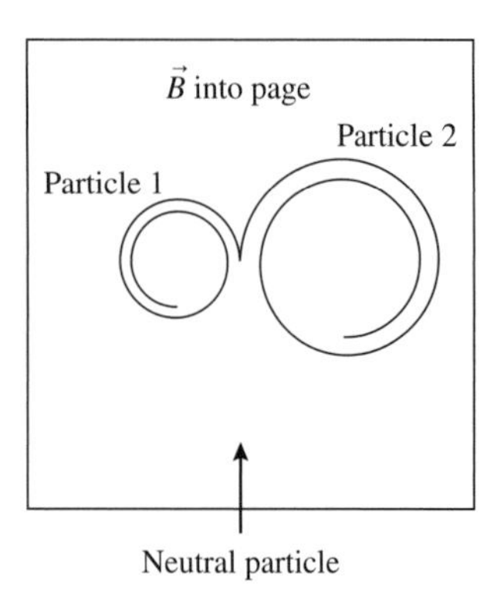

   a. What is the sign (+ or −) of particle 1? ________

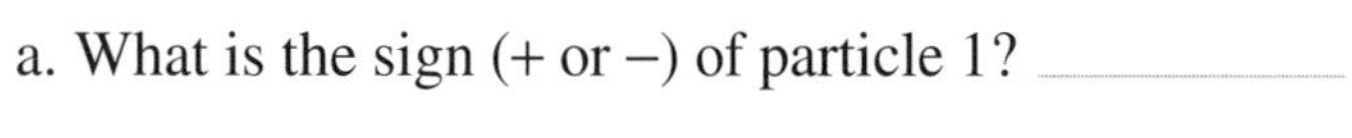

   What is the sign (+ or −) of particle 2? ________

   b. Which charged particle leaves the collision with a larger momentum? Explain. (Assume that $|q| = e$ for both particles.)

22. A solenoid is wound as shown and attached to a battery. Two electrons are fired into the solenoid, one from the end and one through a very small hole in the side.

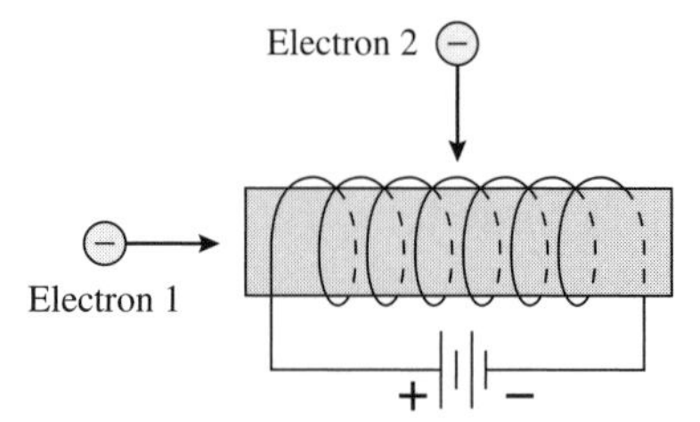

   a. In what direction does the magnetic field inside the solenoid point? Show it on the figure.

   b. Is electron 1 deflected as it moves through the solenoid? If so, in which direction? If not, why not?

   c. Is electron 2 deflected as it moves through the solenoid? If so, in which direction? If not, why not?

24.6 Magnetic Fields Exert Forces on Currents

24.7 Magnetic Fields Exert Torques on Dipoles

23. Three current-carrying wires are perpendicular to the page. Construct a force vector diagram on the figure to find the net force on the upper wire due to the two lower wires.

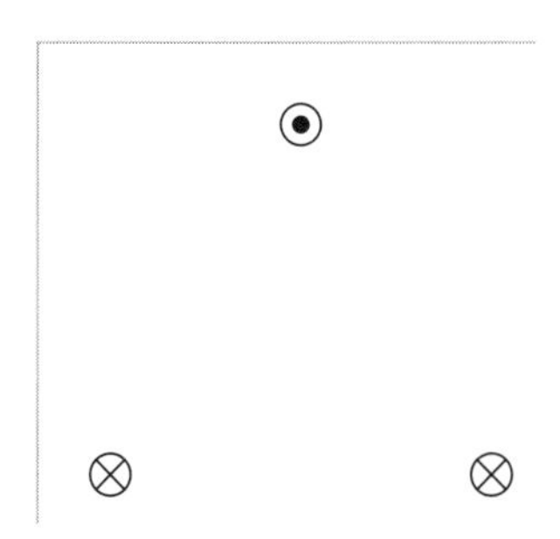

24. Three current-carrying wires are perpendicular to the page.
    a. Construct a force vector diagram on each wire to determine the direction of the net force on each wire.
    b. Can three *charges* be placed in a triangular pattern so that their force diagram looks like this? If so, draw it. If not, why not?

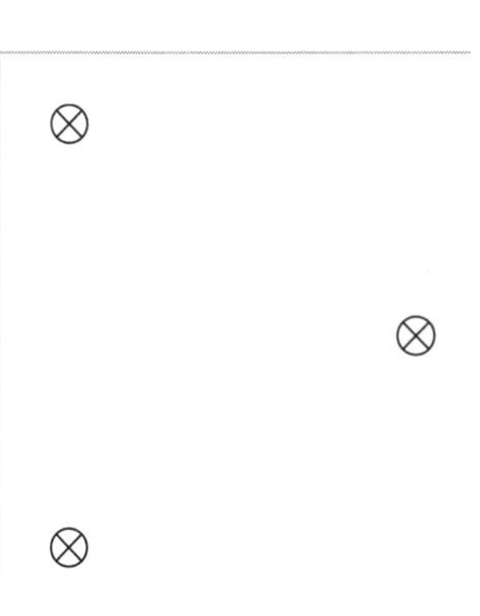

25. A current-carrying wire passes between two bar magnets. Is there a force on the wire? If so, draw the force vector. If not, why not?

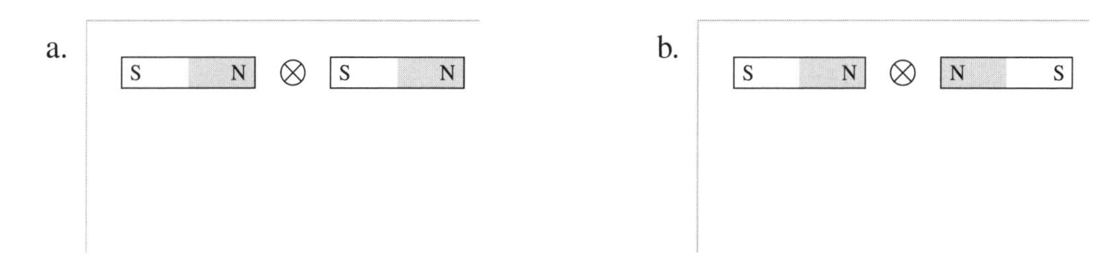

26. A current-carrying wire passes in front of a solenoid that is wound as shown. The wire experiences an upward force.
    a. In what direction does the magnetic field of the solenoid point? Draw the solenoid's magnetic field on the figure.
    b. Use arrows to show the direction in which the current enters and leaves the solenoid. Explain your choice.

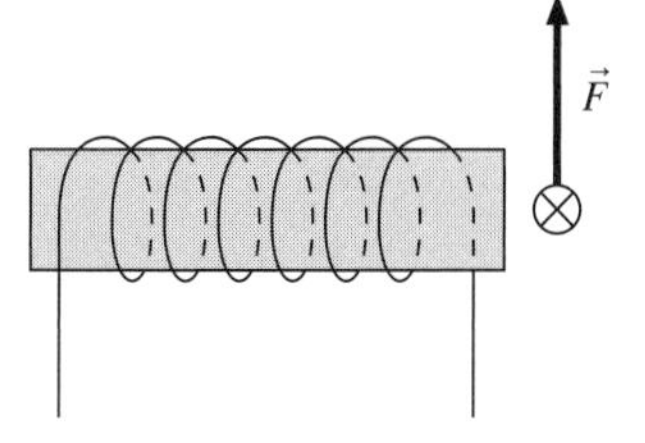

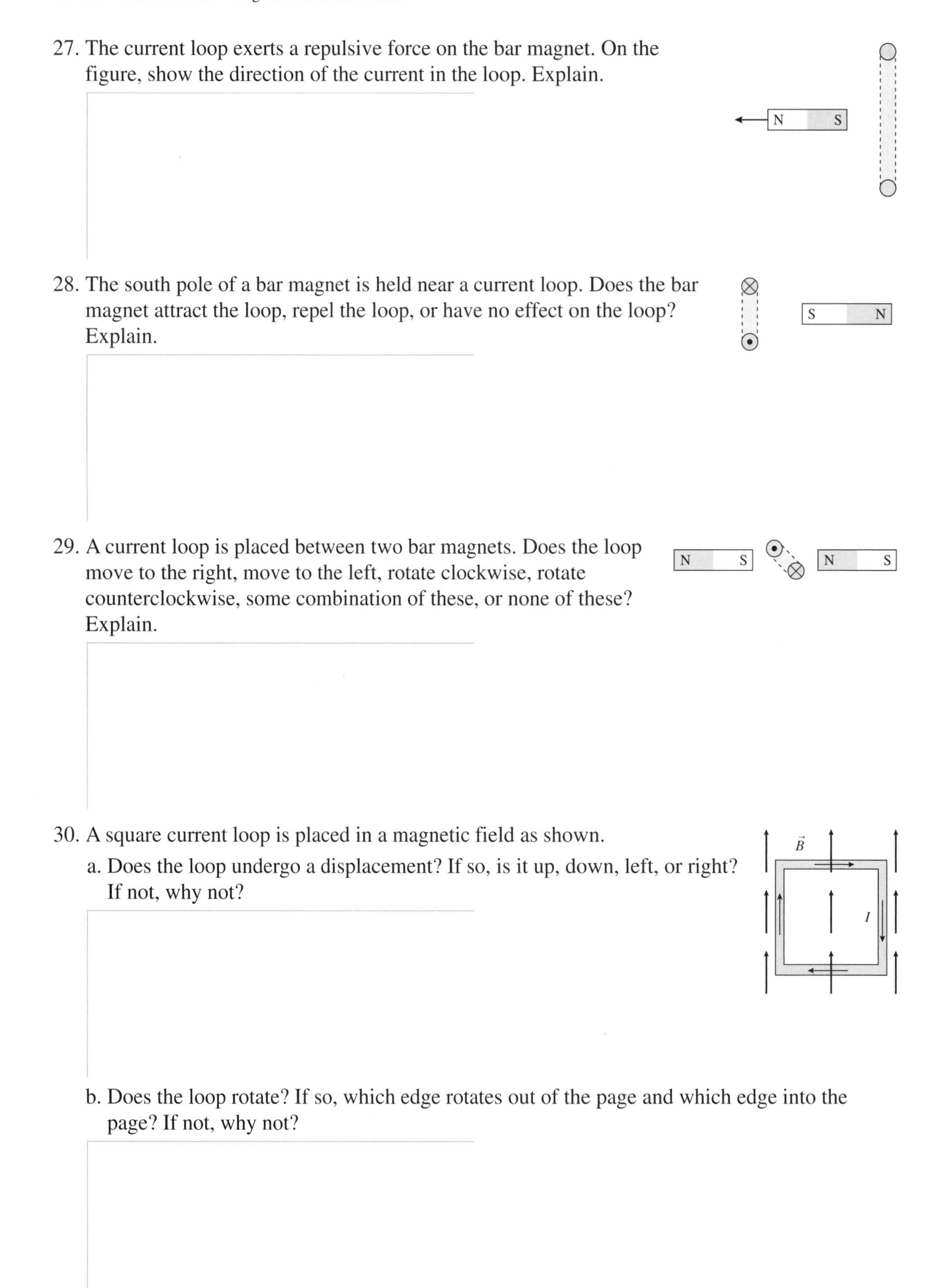

27. The current loop exerts a repulsive force on the bar magnet. On the figure, show the direction of the current in the loop. Explain.

28. The south pole of a bar magnet is held near a current loop. Does the bar magnet attract the loop, repel the loop, or have no effect on the loop? Explain.

29. A current loop is placed between two bar magnets. Does the loop move to the right, move to the left, rotate clockwise, rotate counterclockwise, some combination of these, or none of these? Explain.

30. A square current loop is placed in a magnetic field as shown.
    a. Does the loop undergo a displacement? If so, is it up, down, left, or right? If not, why not?

    b. Does the loop rotate? If so, which edge rotates out of the page and which edge into the page? If not, why not?

24.8 Magnets and Magnetic Materials

31. A solenoid, wound as shown, is placed next to an unmagnetized piece of iron. Then the switch is closed.

Solenoid

Iron

\+ −

a. Identify on the figure the north and south poles of the solenoid.

b. What is the direction of the solenoid's magnetic field as it passes through the iron?

c. What is the direction of the induced magnetic dipole in the iron?

d. Identify on the figure the north and south poles of the induced magnetic dipole in the iron.

e. When the switch is closed, does the iron move left or right? Does it rotate? Explain.

f. Suppose the iron is replaced by a piece of copper. What happens to the copper when the switch is closed?

25 Electromagnetic Induction and Electromagnetic Waves

25.1 Induced Currents

25.2 Motional emf

1. A loop of copper wire is being pulled from between two magnetic poles.
   a. Show on the figure the current induced in the loop. Explain your reasoning.

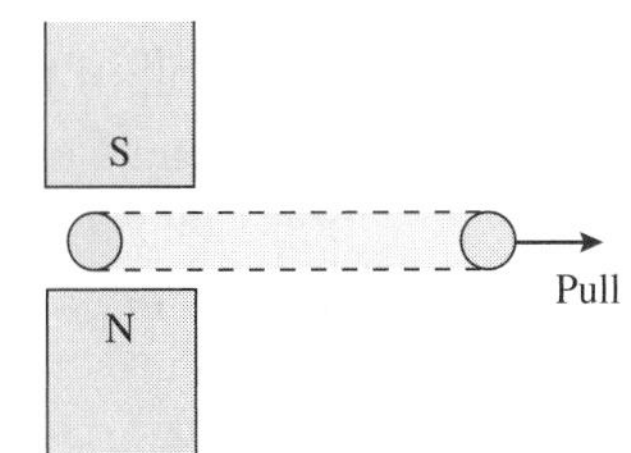

   b. Does either side of the loop experience a magnetic force? If so, draw a vector arrow or arrows on the figure to show any forces.
   c. Label the magnetic poles of the induced current in the loop. Do this on the figure.
   d. Are the magnetic poles you labeled in part c attracted to or repelled by the permanent magnet?

2. A vertical, rectangular loop of copper wire is half in and half out of a horizontal magnetic field (shaded gray). (The field is zero beneath the dotted line.) The loop is released and starts to fall.
   a. Add arrows to the figure to show the direction of the induced current in the loop.
   b. Is there a net magnetic force on the loop? If so, in which direction? Explain.

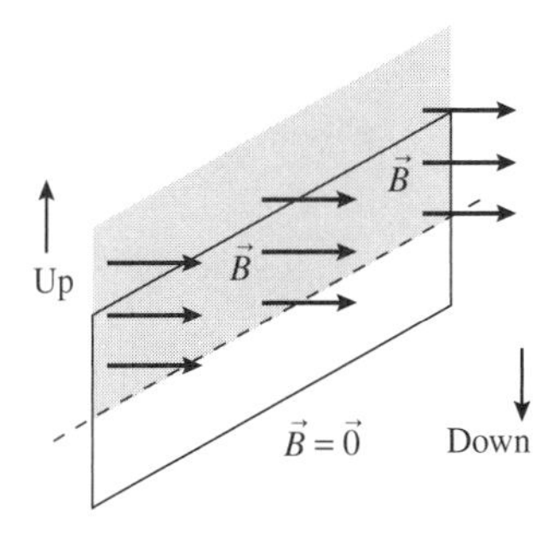

3. An insulating rod pushes a copper loop back and forth. The left edge of the loop, which is always in the magnetic field, oscillates between $x = -L$ and $x = +L$, as shown in the top graph. The right edge of the loop, which includes a lightbulb, is always outside the magnetic field.

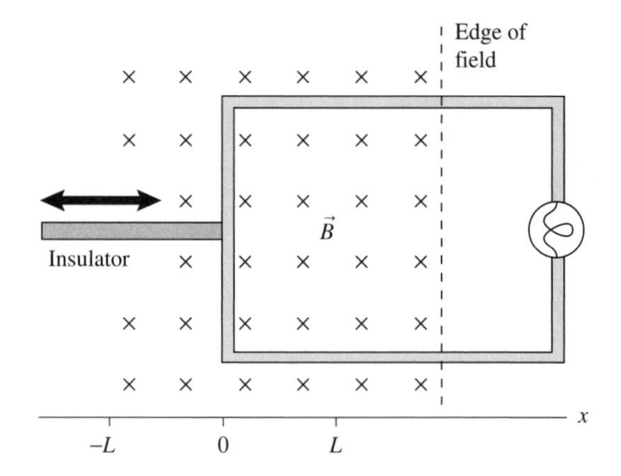

a. Draw the velocity graph for the loop. Make sure it aligns with the position graph above it.

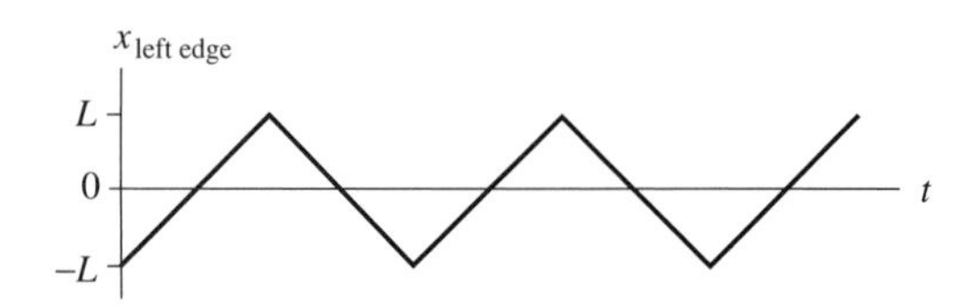

b. Draw a graph of the induced current in the loop as a function of time. Let a clockwise current be a positive number and a counterclockwise current be a negative number.

c. Draw a graph of the brightness of the lightbulb as a function of time.

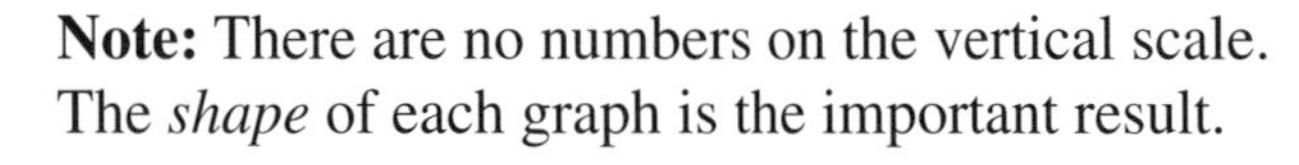

Note: There are no numbers on the vertical scale. The *shape* of each graph is the important result.

25.3 Magnetic Flux

25.4 Faraday's Law

4. A magnetic field is perpendicular to a loop. The graph shows how the magnetic field changes as a function of time, with positive values for B indicating a field into the page and negative values a field out of the page. Several points on the graph are labeled.

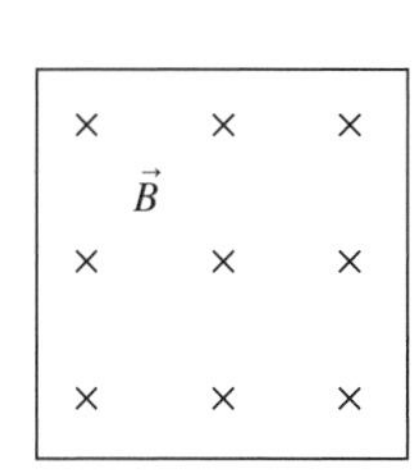

Field through loop

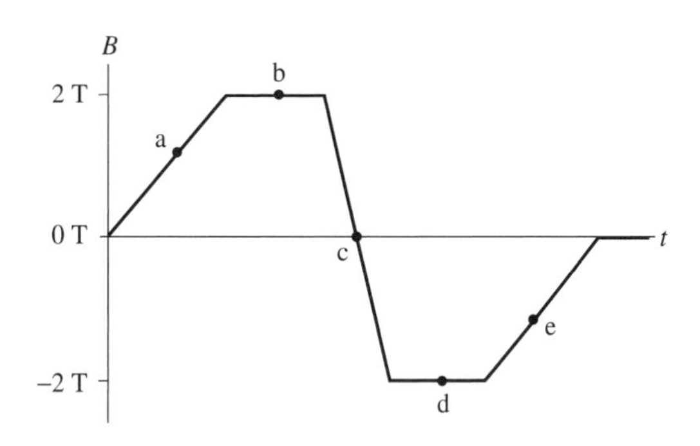

a. At which lettered point or points is the flux through the loop a maximum? ______

b. At which lettered point or points is the flux through the loop a minimum? ______

c. At which point or points is the flux changing most rapidly? ______

d. At which point or points is the flux changing least rapidly? ______

5. Does the loop of wire have a clockwise current, a counterclockwise current, or no current under the following circumstances? Explain.

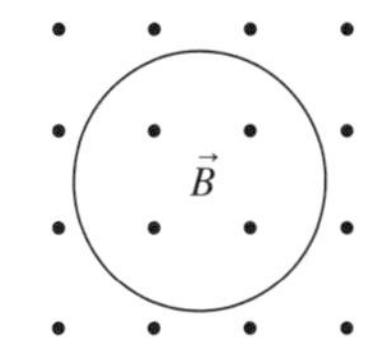

a. The magnetic field points out of the page and its strength is increasing.

b. The magnetic field points out of the page and its strength is constant.

c. The magnetic field points out of the page and its strength is decreasing.

6. A loop of wire is perpendicular to a magnetic field. The magnetic field strength as a function of time is given by the top graph. Draw a graph of the current in the loop as a function of time. Let a positive current represent a current that comes out of the top of the loop and enters the bottom of the loop. There are no numbers for the vertical axis, but your graph should have the correct shape and proportions.

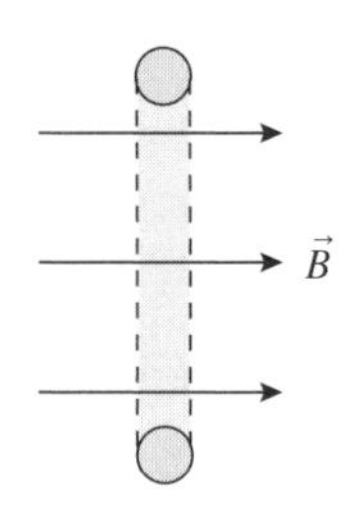

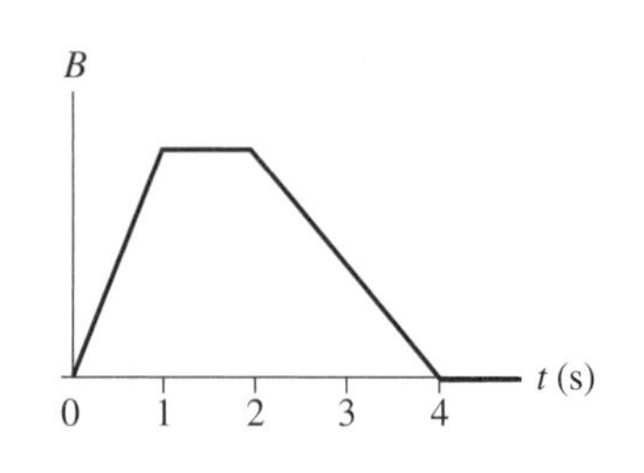

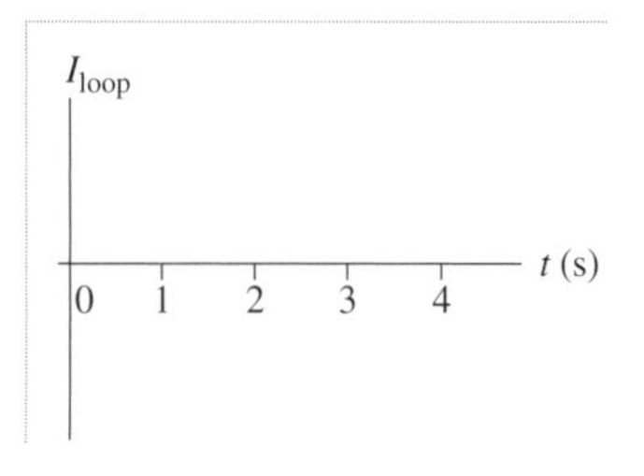

7. a. As the magnet is inserted into the coil, does current flow right to left or left to right through the current meter? Or is the current zero? Explain.

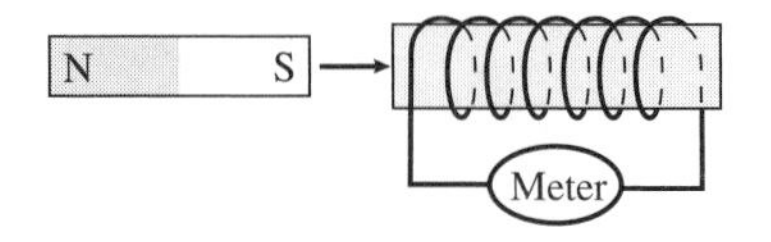

b. As the magnet is held at rest inside the coil, does current flow right to left or left to right through the current meter? Or is the current zero? Explain.

c. As the magnet is withdrawn from the coil, does current flow right to left or left to right through the current meter? Or is the current zero? Explain.

d. If the magnet is inserted into the coil more rapidly than in part a, does the size of the current increase, decrease, or remain the same? Explain.

8. A bar magnet is dropped, south pole down, through the center of a loop of wire. The center of the magnet passes the plane of the loop at time t_c.
   a. Sketch a graph of the magnetic flux through the loop as a function of time.
   b. Sketch a graph of the current in the loop as a function of time. Let a clockwise current be a positive number and a counterclockwise current be a negative number.

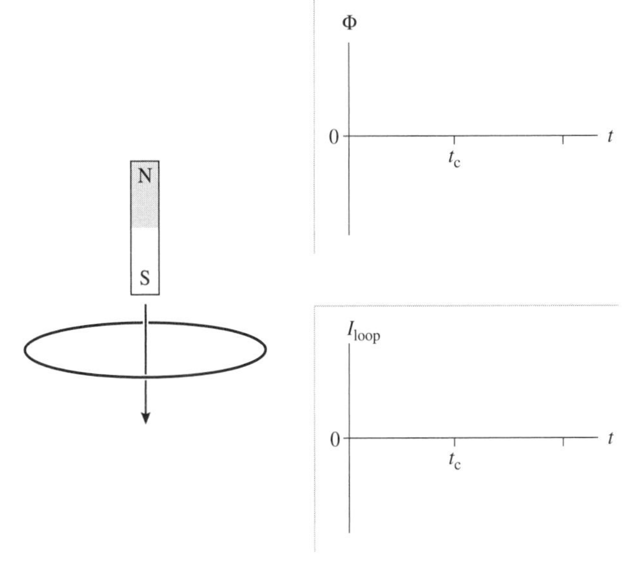

9. a. Just after the switch on the left coil is closed, does current flow right to left or left to right through the current meter of the right coil? Or is the current zero? Explain.

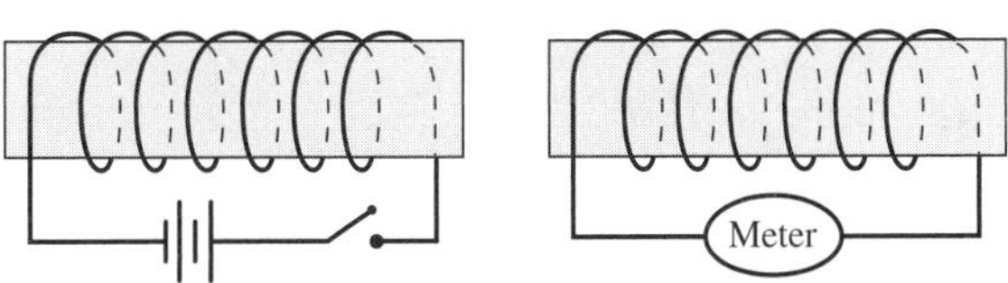

   b. Long after the switch on the left coil is closed, does current flow right to left or left to right through the current meter of the right coil? Or is the current zero? Explain.

   c. Just after the switch on the left coil is reopened, does current flow right to left or left to right through the current meter of the right coil? Or is the current zero? Explain.

10. A solenoid is perpendicular to the page, and its field strength is increasing. Three circular wire loops of equal radii are shown. Rank in order, from largest to smallest, the size of the induced emf in the three rings.

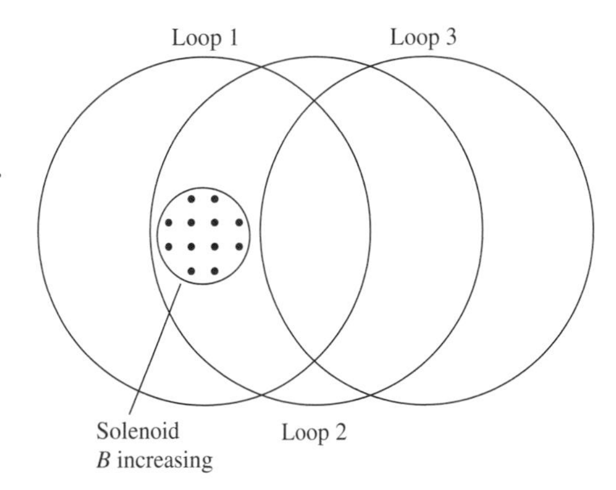

Order:

Explanation:

11. Two very thin sheets of copper are pulled through a magnetic field. Do eddy currents flow in the sheet? If so, show them on the figures, with arrows to indicate the direction of flow. If not, why not?

a.

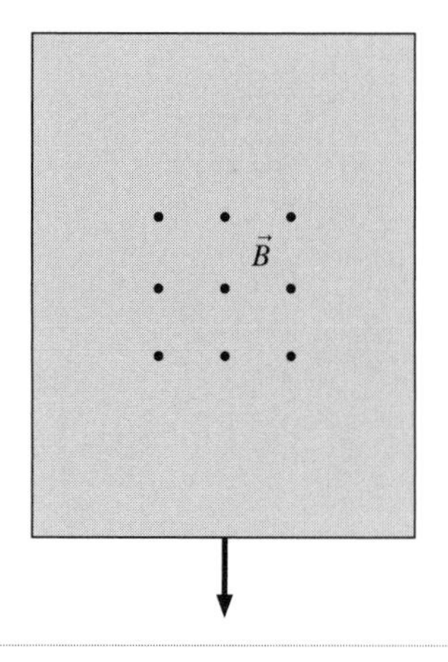

b.

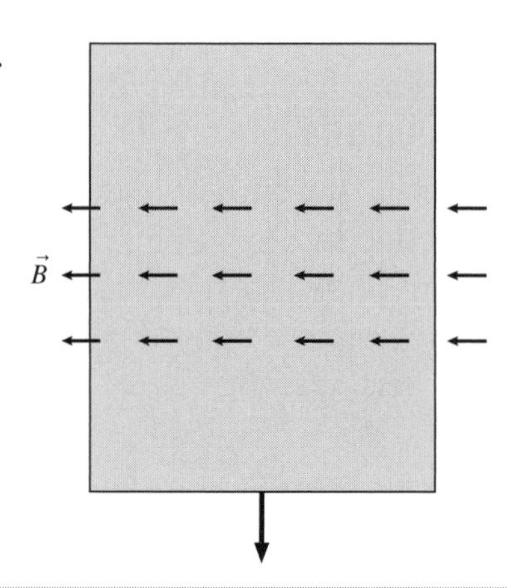

12. The figure shows an edge view of a copper sheet being pulled between two magnetic poles.

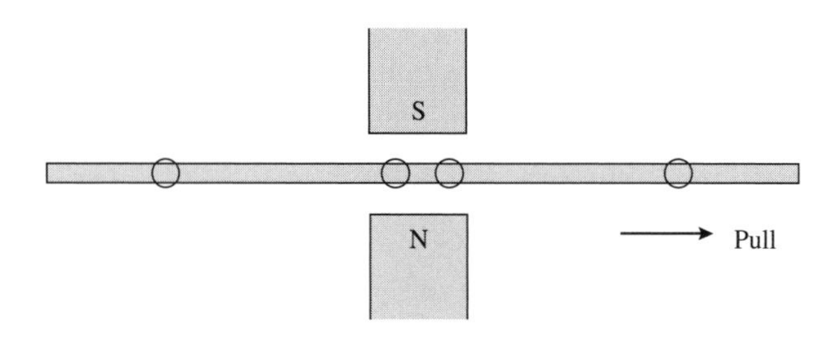

a. Add a dot or an × to each of the circles to indicate the direction in which eddy currents are flowing in and out of the page.

b. Label the magnetic poles of any induced current loops.

c. Do the magnetic poles you labeled in part b experience magnetic forces? If so, add force vectors to the figure to show the directions. If not, why not?

d. Is there a net magnetic force on the copper sheet? If so, in which direction?

25.5 Induced Fields and Electromagnetic Waves

25.6 Properties of Electromagnetic Waves

13. This is an electromagnetic plane wave traveling into the page.
Draw the magnetic field vectors $\vec{B}$ at the dots.

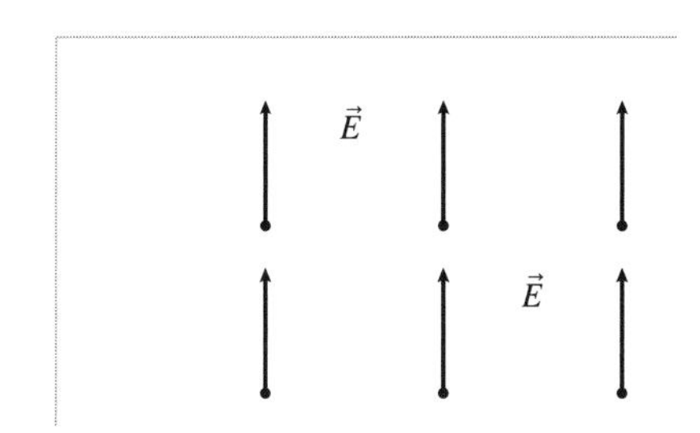

14. This is an electromagnetic wave.

a. Draw the velocity vector $\vec{v}_{em}$.

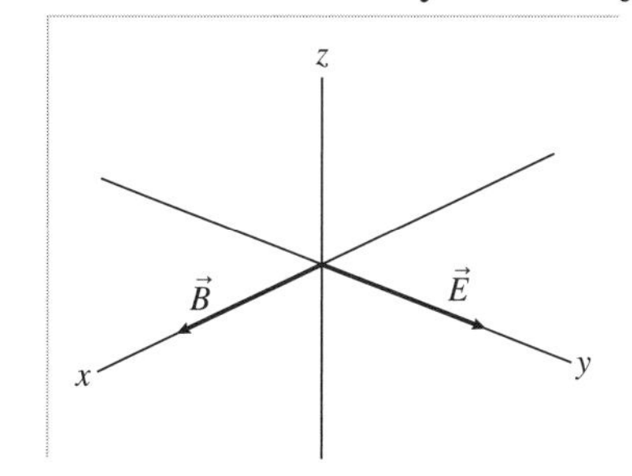

b. Draw $\vec{E}$, $\vec{B}$, and $\vec{v}_{em}$ a half cycle later.

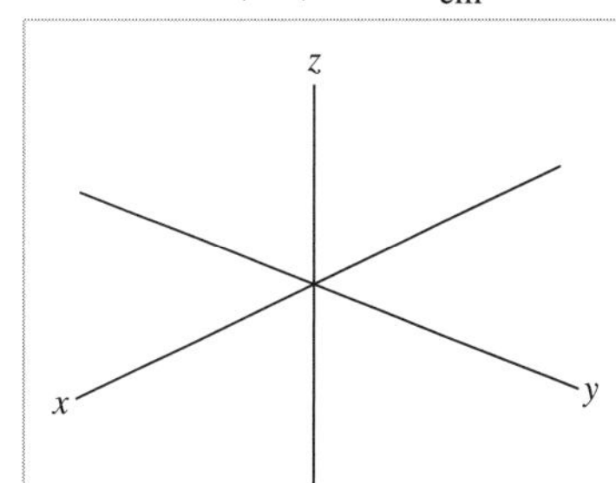

15. Do the following represent possible electromagnetic waves? If not, why not?

a.

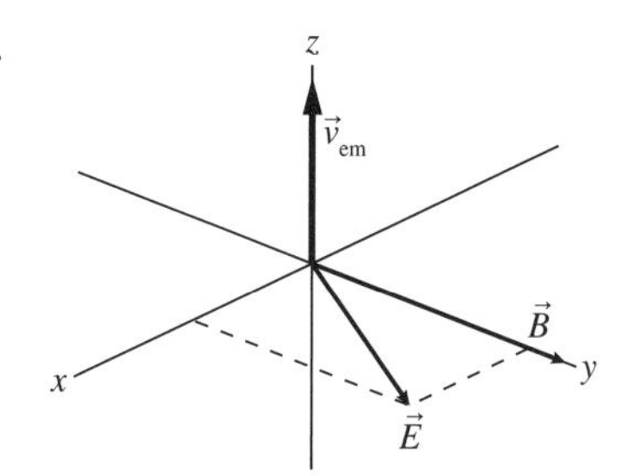

b.

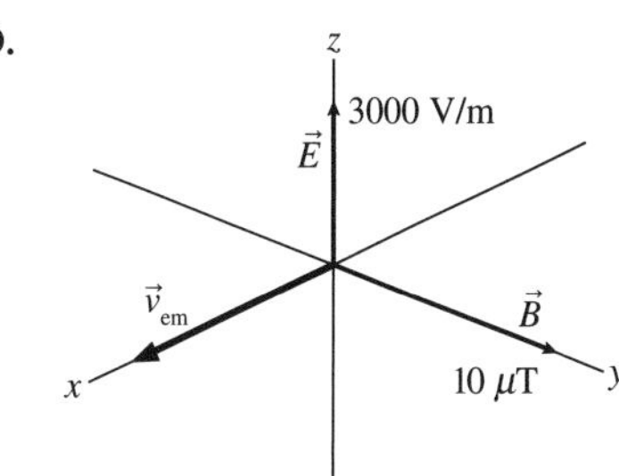

c.

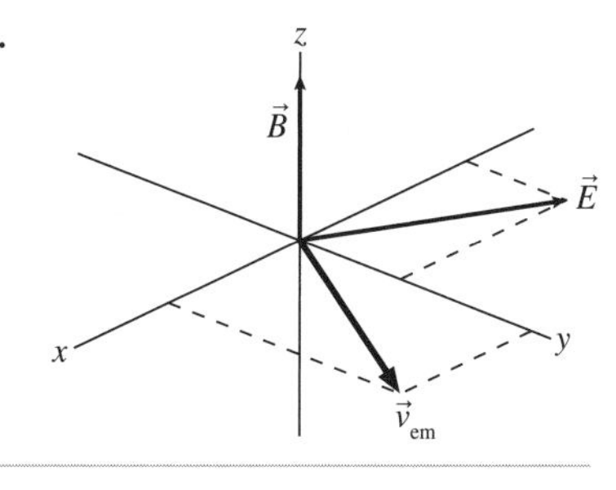

d.

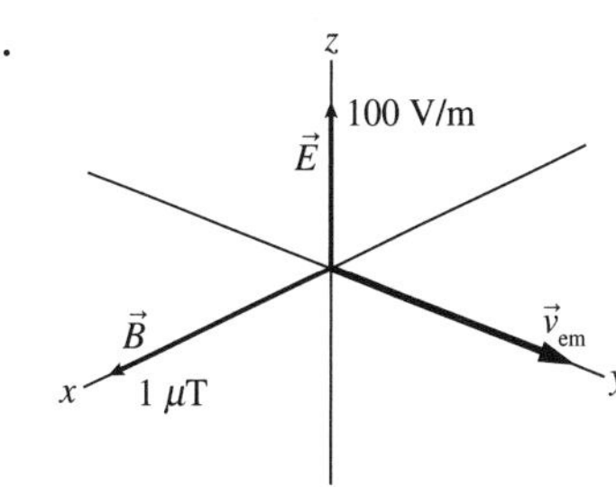

16. The intensity of an electromagnetic wave is 10 W/m^2. What will be the intensity if:
    a. The amplitude of the electric field is doubled?

    b. The amplitude of the magnetic field is doubled?

    c. The amplitudes of both the electric field and the magnetic field are doubled?

    d. The frequency is doubled?

17. A polarized electromagnetic wave passes through a polarizing filter. Draw the electric field of the wave after it has passed through the filter.

a.

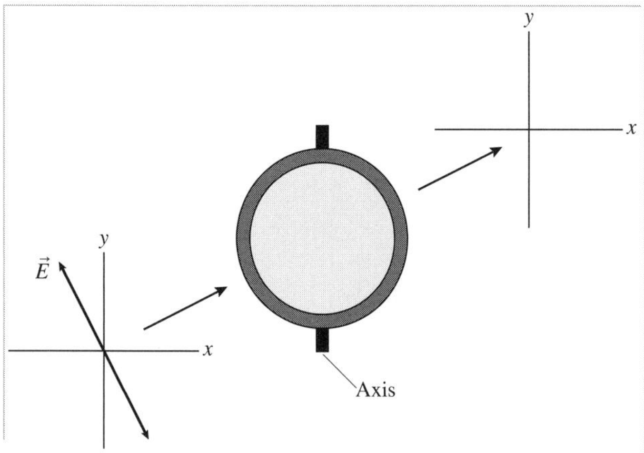

b.

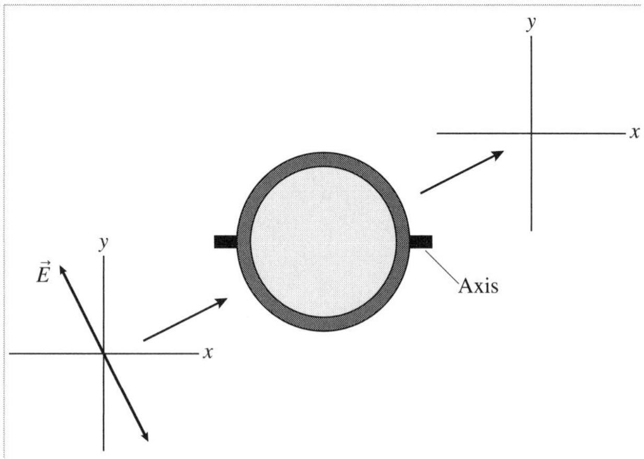

18. A polarized electromagnetic wave passes through a series of polarizing filters. Draw the electric field of the wave after it has passed through each filter.

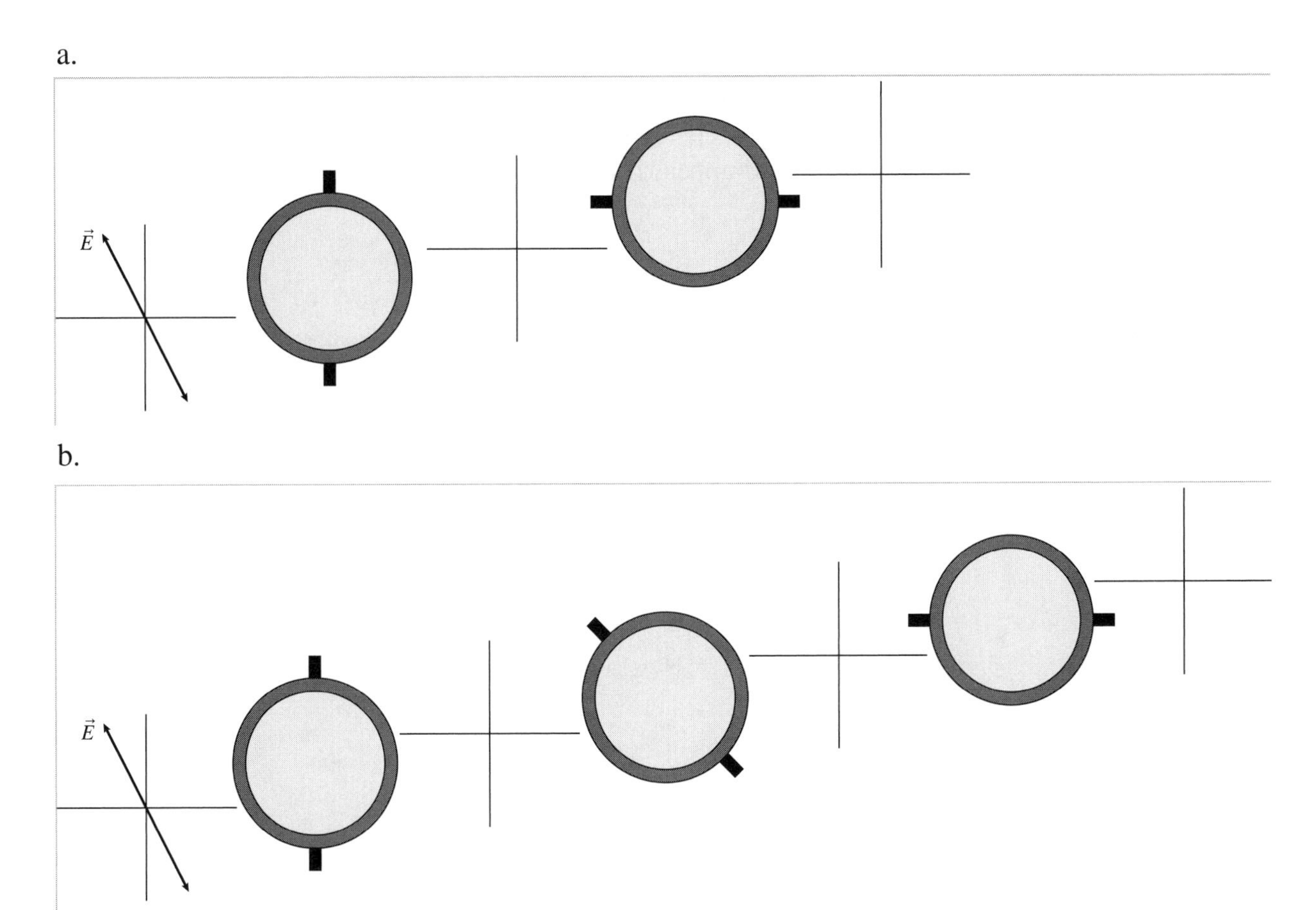

25.7 The Photon Model of Electromagnetic Waves

25.8 The Electromagnetic Spectrum

19. The graph shows the power spectrum of light as a function of wavelength from an object at a temperature *T*. The area under the graph represents the total power output. Sketch on the graph approximately how the spectrum would appear if the object's temperature were doubled to 2*T*.

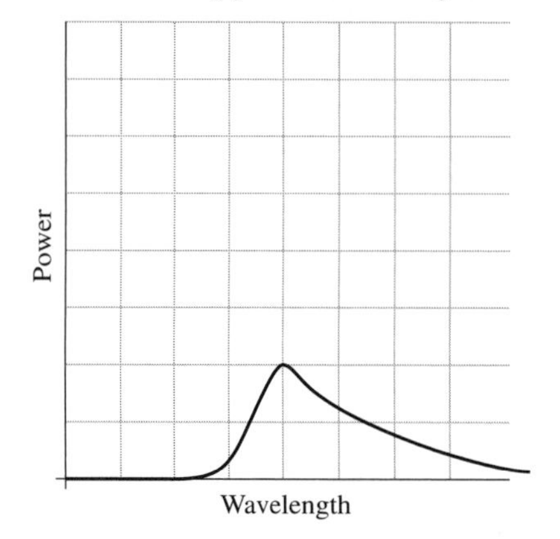

26 AC Circuits

26.1 Alternating Current

1. The graph shows the AC current through a resistor in a simple AC resistor circuit as a function of time for two cycles. This current can be expressed as $I_R = I_0 \cos(2\pi ft)$.

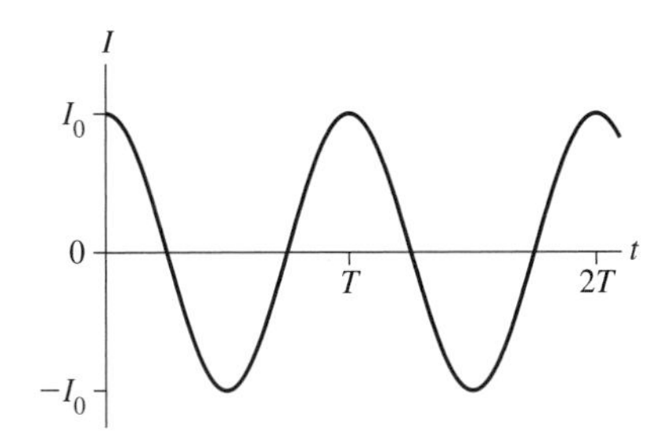

   a. On the axes below the graph, sketch the *square* of the current through the resistor as a function of time. Be careful to make sure that your graph's features are aligned vertically with those of the AC current graph above it.

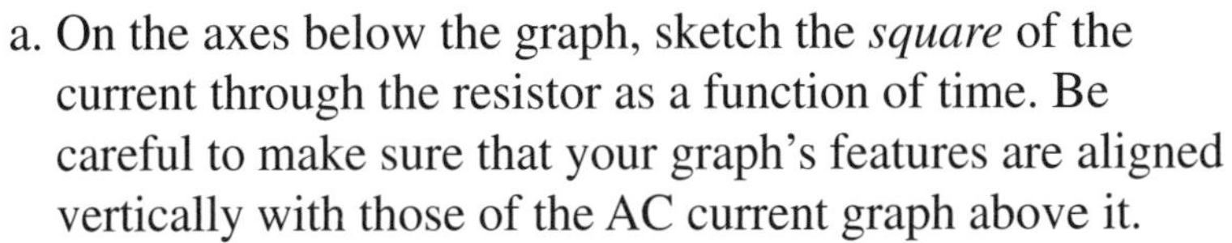

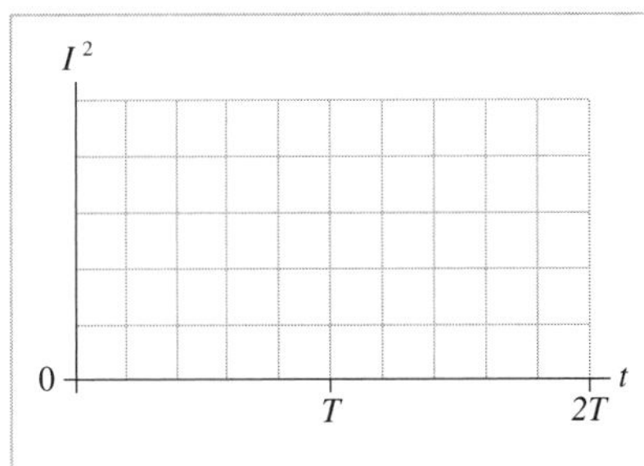

   b. What is the time-average value of the current through the resistor after an integer number of whole cycles? ________

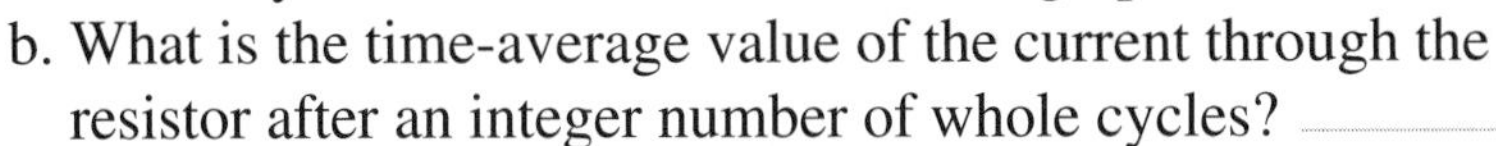

   c. If the frequency of the AC current is f, what is the frequency of the square of the AC current? ________

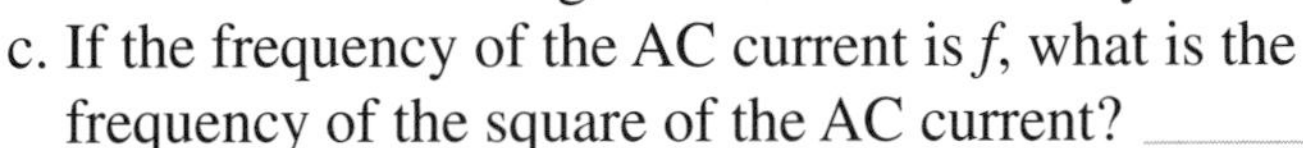

   d. A useful trigonometric identity for the square of a cosine function is $\cos^2\theta = \frac{1}{2} + \frac{1}{2}\cos(2\theta)$. What is the time-average value of the function $\cos(2\pi ft)$ after an integer number of whole cycles? ________

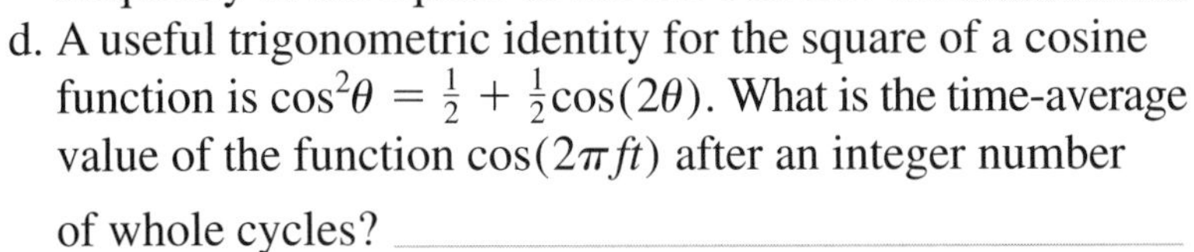

   e. In light of your answer to part d, what is the average value of the square of the current through the resistor after an integer number of whole cycles?

   f. What is the square root of your answer to part e, that is, the square root of the average value of the square? (The value you obtain is what is referred to as the root-mean-square of the current (I_{rms}).)

26.2 The Transmission and Use of Electricity

2. A student has two D-cell (1.5 V) batteries, but needs a 9-V battery to run his transistor radio. He also has a transformer with $N_1 = 1000$ turns and $N_2 = 3000$ turns. He proposes to use the transformer to obtain the 9 V that he needs. Can he do this? If not, why not? Explain.

26.3 Biological Effects and Electrical Safety

3. The figure shows a drawing of a standard (North American) household electrical socket.

a. Label the socket hole that is neutral (usually at ground voltage = 0).
b. When you plug a device into the outlet, does the current always enter through one prong and leave through the other? What does it mean that one prong is connected to neutral? Explain.

26.4 Capacitor Circuits

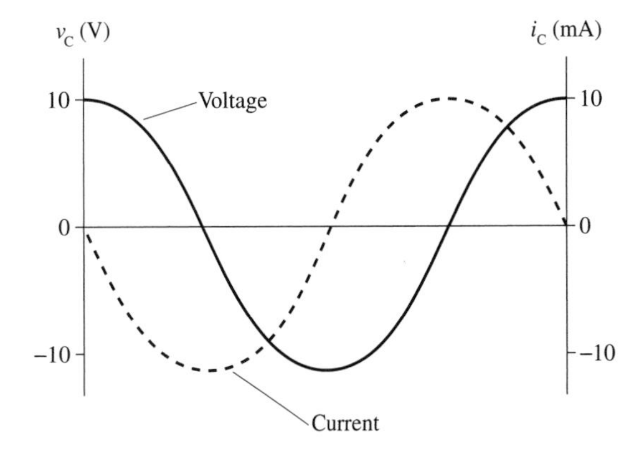

4. Current and voltage graphs are shown for a capacitor circuit with $\omega = 1000$ rad/s.

   a. What is the capacitive reactance X_C?

   b. What is the capacitance C?

5. Consider these three circuits.

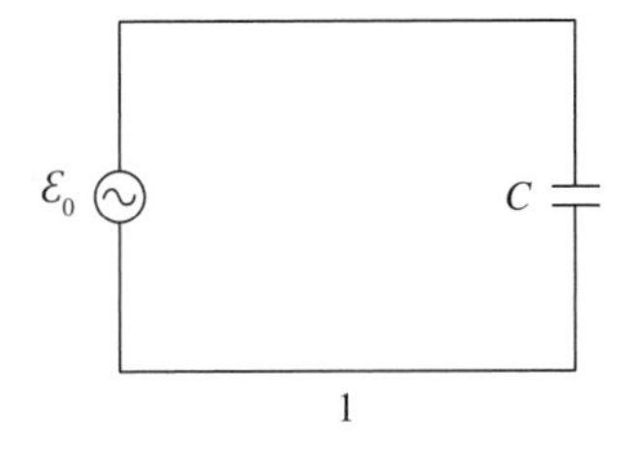

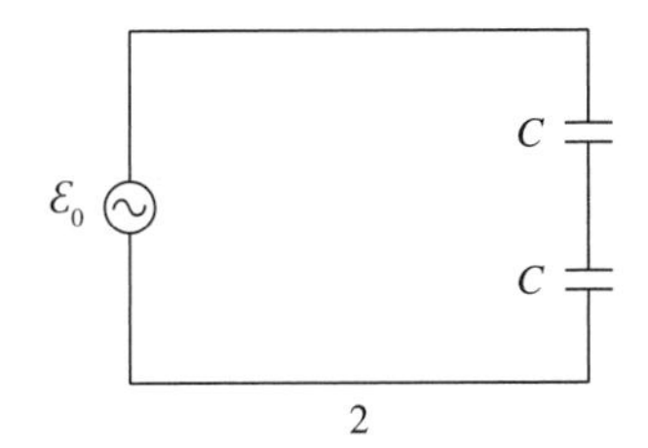

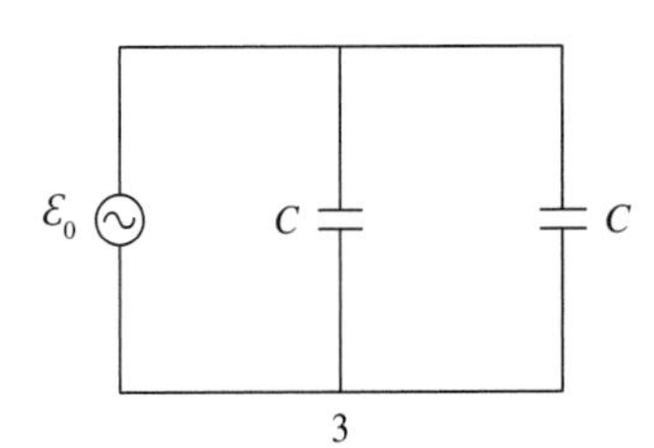

Rank in order, from largest to smallest, the peak currents $(I_C)_1$ to $(I_C)_3$.

Order:

Explanation:

26.5 Inductors and Inductor Circuits

6. Current and voltage graphs are shown for an inductor circuit with $\omega = 1000$ rad/s.

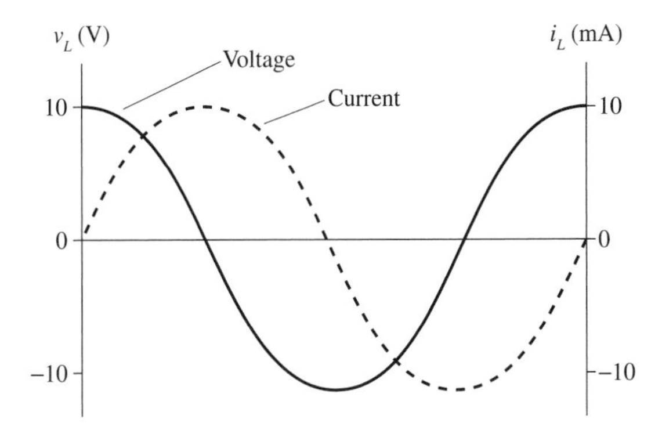

a. What is the inductive reactance X_L?

b. What is the inductance L?

7. The figure shows the current through an inductor. A positive current is defined as a current going from top to bottom. At the time corresponding to each of the labeled points, does the potential across the inductor (going from top to bottom) increase, decrease, or stay the same?

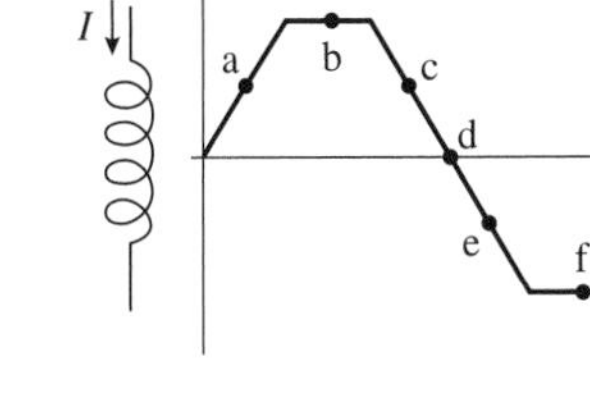

a. ____________ e. ____________

b. ____________ f. ____________

c. ____________ g. ____________

d. ____________

8. Rank in order, from most positive to most negative, the inductor's potential difference $(\Delta V_L)_a$, $(\Delta V_L)_b$, …, $(\Delta V_L)_f$, at the six labeled points. ΔV_L is the change in going from the top of the inductor to the bottom. Some may be equal. Note that 0 V > −2 V.

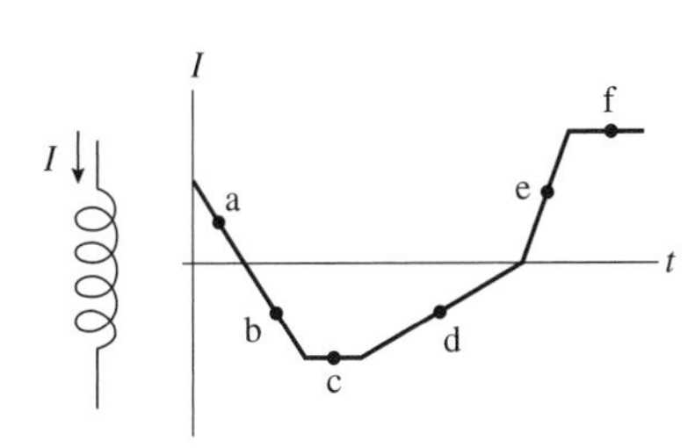

Order:

Explanation:

9. a. Can you tell which of these inductors has the larger current flowing through it? If so, which one? If not, why not?

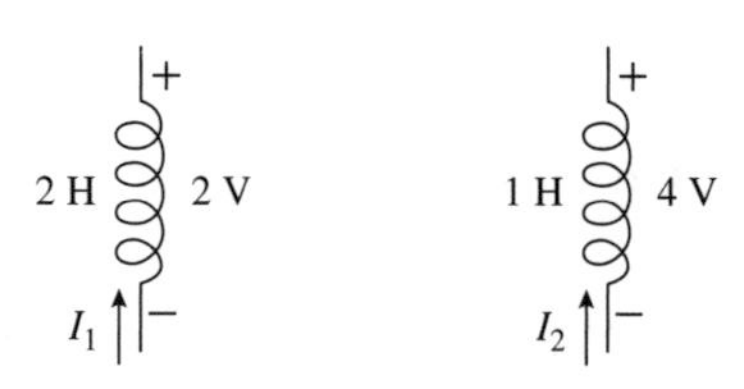

b. Can you tell through which inductor the current is changing most rapidly? If so, which one? If not, why not?

c. If the current enters the inductor from the bottom, can you tell if the current is increasing, decreasing, or staying the same? If so, which one and what is your reasoning? If not, why not?

26.6 Oscillation Circuits

10. An *LC* circuit oscillates at a frequency of 2000 Hz. What will the frequency be if the inductance is quadrupled?

11. Three *LC* circuits are made with the same capacitor but different inductors. The figure shows the inductor current as a function of time. Rank in order, from largest to smallest, the three inductances L_1, L_2, and L_3. Some may be equal.

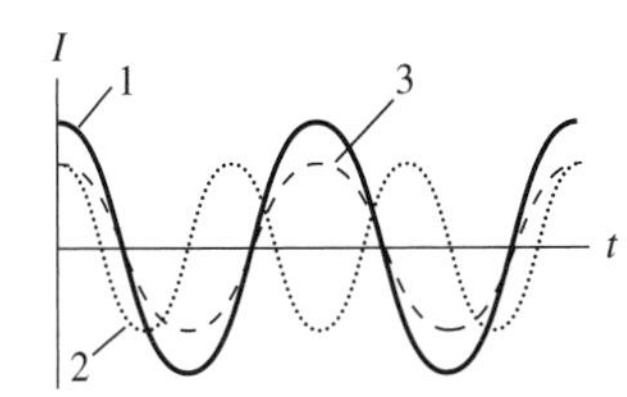

Order:

Explanation:

12. An *RLC* circuit has an inductance of 1 mH, capacitance of 1 mF so that the natural frequency of the circuit is about 160 Hz. The circuit is driven by a variable-frequency AC source with a peak voltage of 20 V.

    a. On the axes below, draw and label graphs of the peak current amplitude as a function of the frequency of the AC source voltage when the circuit has a total resistance of 2.0 Ω and 20 Ω, respectively.

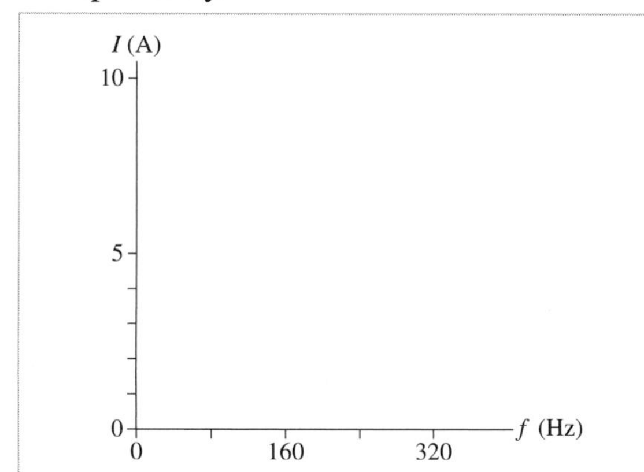

    b. Compare your results above to those of workbook exercise 14.18 if you previously completed that exercise.

27 Relativity

27.1 Relativity: What's It All About?

27.2 Galilean Relativity

1. In which reference frame, S or S′, does the ball move faster?

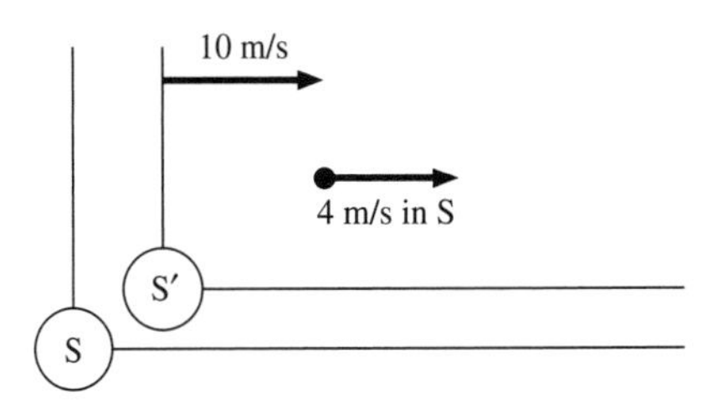

2. Frame S′ moves parallel to the x-axis of frame S.
   a. Is there a value of v for which the ball is at rest in S′? If so, what is v? If not, why not?

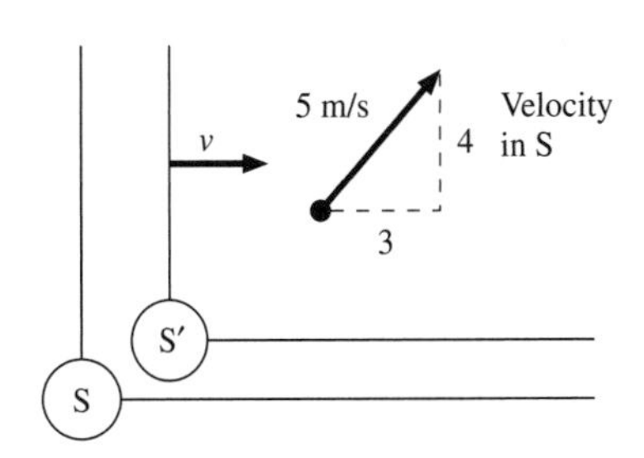

   b. Is there a value of v for which the ball has a minimum speed in S′? If so, what is v? If not, why not?

3. Anjay can swim at a steady speed of 2 mph. He needs to cross a river that flows west to east at 4 mph. Anjay jumps in at point A and swims due north (i.e., his head always points due north) until reaching the opposite shore. Where does Anjay land?

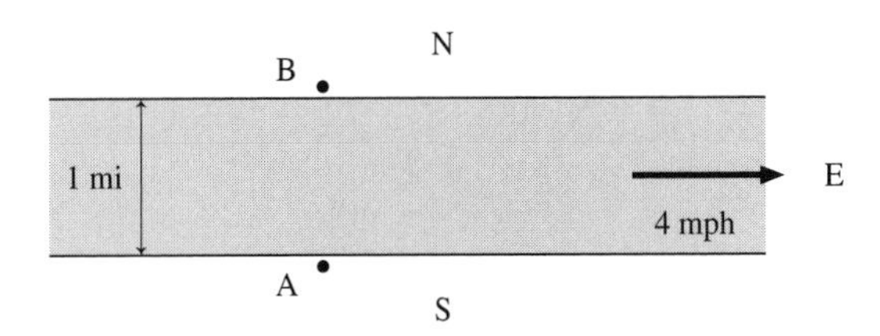

4. What are the speed and direction of each ball in a reference frame that moves to the right at 2 m/s?

5. a. The graph below left shows the positions of Al's car (A) and Bob's car (B) traveling along a highway. On the axes to the right, draw a graph of the position and velocity of Al's car as measured from Bob's reference frame. Indicate numerical values on the position and velocity axes.

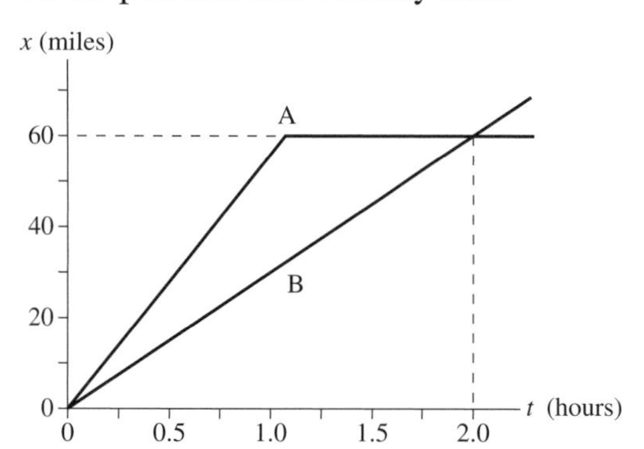

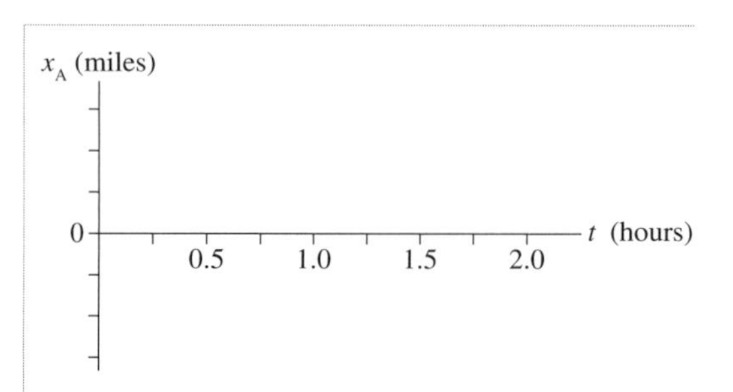

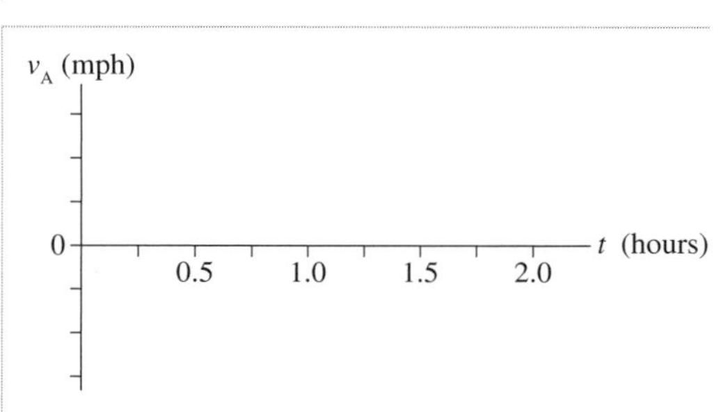

b. The graph below left shows the position of Donna's car (D) as measured from Bob's reference frame. On the axes to the right, show the position and velocity of Donna's car with respect to the highway. Indicate numerical values on the position and velocity axes.

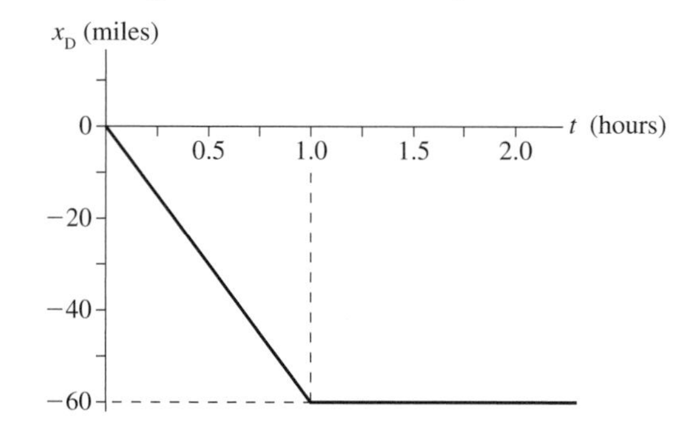

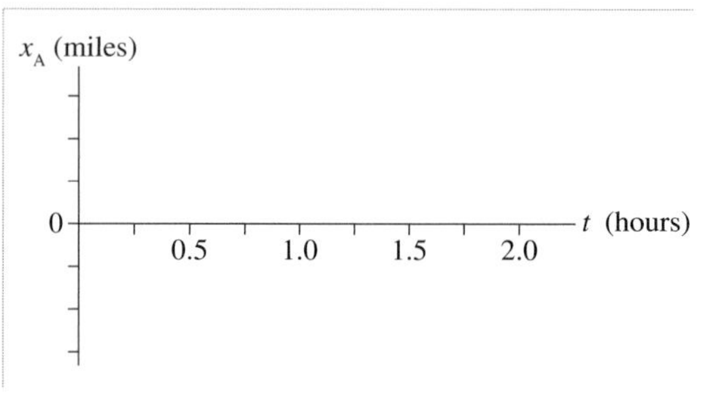

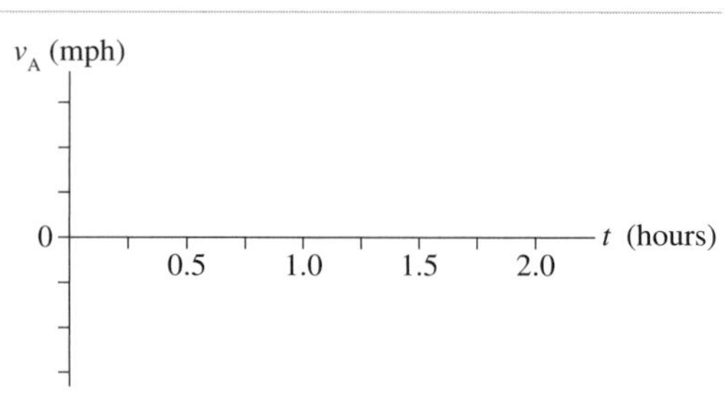

27.3 Einstein's Principle of Relativity

6. Teenagers Sam and Tom are playing chicken in their rockets. As seen from the earth, each is traveling at $0.95c$ as he approaches the other. Sam fires a laser beam toward Tom.

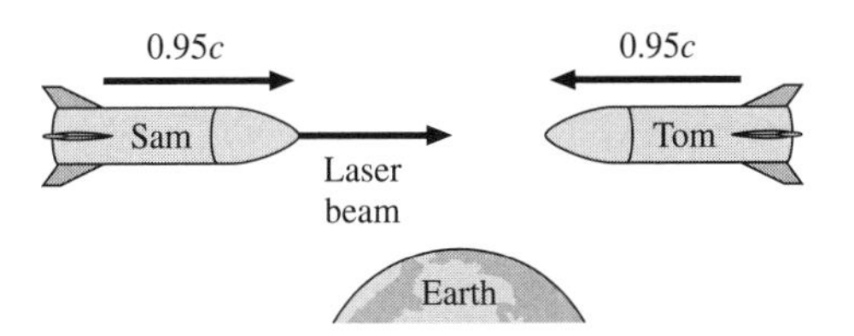

   a. What is the speed of the laser beam relative to Sam?

   b. What is the speed of the laser beam relative to Tom?

27.4 Events and Measurements

7. It is a bitter cold day at the South Pole, so cold that the speed of sound is only 300 m/s. The speed of light, as always, is 300 m/μs. A firecracker explodes 600 m away from you.

   a. How long after the explosion until you see the flash of light? ______

   b. How long after the explosion until you hear the sound? ______

   c. Suppose you see the flash at $t = 2.000002$ s. At what time was the explosion? ______

   d. What are the spacetime coordinates for the event "firecracker explodes"? Assume that you are at the origin and that the explosion takes place at a position on the positive x-axis.

8. You are at the origin of a coordinate system containing clocks, but you're not sure if the clocks have been synchronized. The clocks have reflective faces, allowing you to read them by shining light on them. You flash a bright light at the origin at the instant your clock reads $t = 2.000000$ s.

   a. At what time will you see the reflection of the light from a clock at $x = 3000$ m?

   b. When you see the clock at $x = 3000$ m, it reads 2.000020 s. Is the clock synchronized with your clock at the origin? Explain.

9. Can two spatially separated events be simultaneous if they are seen at two different times? If not, why not? If so, give an example.

10. Can two simultaneous events, A and B, at different locations be seen by different observers as taking place in a different order, so that one observer sees A then B, but another observer sees B followed by A? If not, why not? If so, give an example.

11. Can two events occurring at the same location be seen as simultaneous by an observer at one place, but not by an observer in another place? If not, why not? If so, give an example.

27.5 The Relativity of Simultaneity

12. Two supernovas, labeled L and R, occur on opposite sides of a galaxy, at equal distances from the center. The supernovas are seen at the same instant on a planet at rest in the center of the galaxy. A spaceship is entering the galaxy from the left at a speed of 0.999c relative to the galaxy.

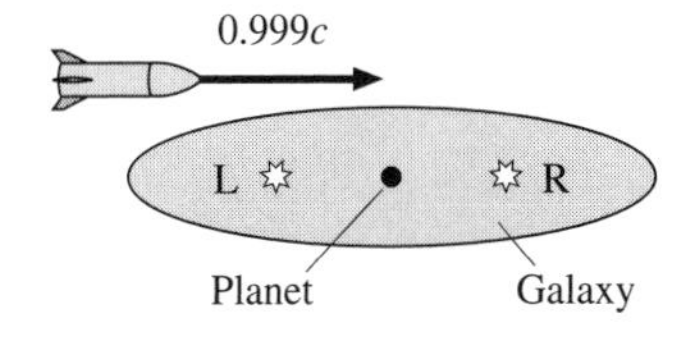

a. According to astronomers on the planet, were the two explosions simultaneous? Explain why.

b. Which supernova, L or R, does the spaceship crew *see* first? ______

c. Did the supernova that was *seen* first necessarily *happen* first in the rocket's frame? Explain.

d. Is "two light flashes reach the planet at the same instant" an event? To help you decide, could you arrange for something to happen only if two light flashes from opposite directions arrive at the same time? Explain.

If you answered Yes to part d, then the crew on the spaceship will also determine, from their measurements, that the light flashes reach the planet at the same instant. (Experimenters in different reference frames may disagree about when and where an event occurs, but they all agree that it *does* occur.)

e. The figure below shows the supernovas in the spaceship's reference frame with the *assumption* that the supernovas are simultaneous. The second half of the figure is a short time after the explosions. Draw two circular wave fronts to show the light from each supernova. Neither wave front has yet reached the planet. Be sure to consider:

- The points on which the wave fronts are centered.
- The wave speeds of each wave front.

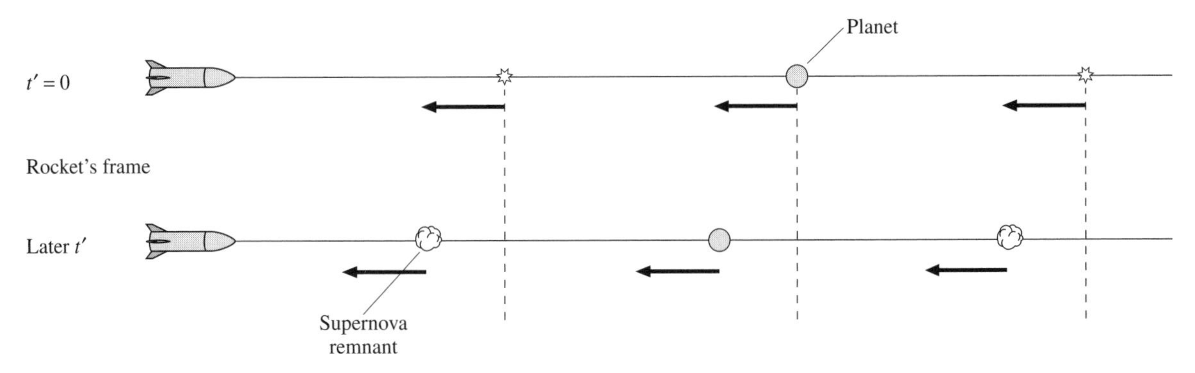

f. According to your diagram, are the two wave fronts going to reach the planet at the same instant of time? Why or why not?

g. Does your answer to part f conflict with your answer to part d?

If so, what different assumption could you make about the supernovas in the rocket's frame that would bring your wave-front diagram into agreement with your answer to part d?

h. So according to the spaceship crew, are the two supernovas simultaneous? If not, which happens first?

13. A rocket is traveling from left to right. At the instant it is halfway between two trees, lightning simultaneously (in the rocket's frame) hits both trees.

a. Do the light flashes reach the rocket pilot simultaneously? If not, which reaches him first? Explain.

b. A student was sitting on the ground halfway between the trees as the rocket passed overhead. According to the student, were the lightning strikes simultaneous? If not, which tree was hit first? Explain.

27.6 Time Dilation

14. Your friend flies from Los Angeles to New York. She carries an accurate stopwatch with her to measure the flight time. You and your assistants on the ground also measure the flight time.
    a. Identify the two events associated with this measurement.

    b. Who, if anyone, measures the proper time?

    c. Who, if anyone, measures the shorter flight time?

    d. Who, if anyone, measures the longer flight time?

15. Evaluate the quantity $\dfrac{1}{\sqrt{1-\beta^2}}$ for the following relative velocities

    a. $v = 3/5\ c$

    b. $v = 4/5\ c$

    c. $v = 5/13\ c$

    d. $v = 12/13\ c$

    e. How does the quantity $\dfrac{1}{\sqrt{1-\beta^2}}$ change with increasing v? Does a factor of two increase in v lead to a greater or lesser increase in $\dfrac{1}{\sqrt{1-\beta^2}}$?

27.7 Length Contraction

16. Experimenters in B's reference frame measure $L_A = L_B$. Do experimenters in A's reference frame agree that A and B are the same length? If not, which do they find to be longer? Explain.

A v

B

17. As a meter stick flies past you, you simultaneously measure the positions of both ends and determine that $L < 1$ m.

Meter stick
v

a. To an experimenter in frame S′, the meter stick's frame, did you make your two measurements simultaneously? If not, which end did you measure first? Explain.

Hint: Review the reasoning about simultaneity that you used in Exercises 12–13.

b. Can experimenters in frame S′ give an explanation for why your measurement is < 1 m?

c. Is there a reference frame in which the meter stick is measured to have a length greater than one meter? Explain.

27.8 Velocities of Objects in Special Relativity

18. A rocket travels at speed $0.5c$ relative to the earth.

a. The rocket shoots a bullet in the forward direction at speed $0.5c$ relative to the rocket. Is the bullet's speed relative to the earth less than, greater than, or equal to c?

Earth
$0.5c$

b. The rocket shoots a second bullet in the backward direction at speed $0.5c$ relative to the rocket. In the earth's frame, is the bullet moving right, moving left, or at rest?

27.9 Relativistic Momentum

19. Particle A has half the mass and twice the speed of particle B. Is p_A less than, greater than, or equal to p_B? Explain.

20. Particle A has one-third the mass of particle B. The two particles have equal momenta. Is u_A less than, greater than, or equal to $3u_B$? Explain.

21. Event B occurs at $t_B = 10.0\ \mu s$. An earlier event A, at $t_A = 5.0\ \mu s$, is the cause of B. What is the maximum possible distance that A can be from B?

27.10 Relativistic Energy

22. Consider these 4 particles:

Particle	Rest energy	Total energy
1	A	A
2	B	$2B$
3	$2C$	$4C$
4	$3D$	$5D$

Rank in order, from largest to smallest, the particles' speeds u_1 to u_4.

Order:

Explanation:

28 Quantum Physics

28.1 X Rays and X-Ray Diffraction

1. Use trigonometry to show from the diagram that the two rays shown will constructively interfere when the Bragg condition is met.

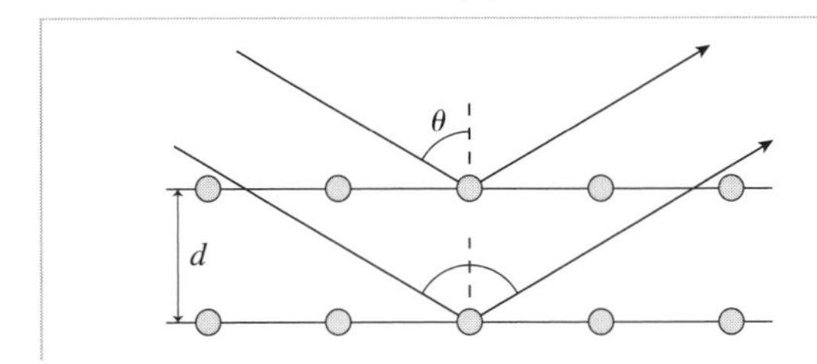

28.2 The Photoelectric Effect

2. a. A negatively charged electroscope can be discharged by shining an ultraviolet light on it. How does this happen?

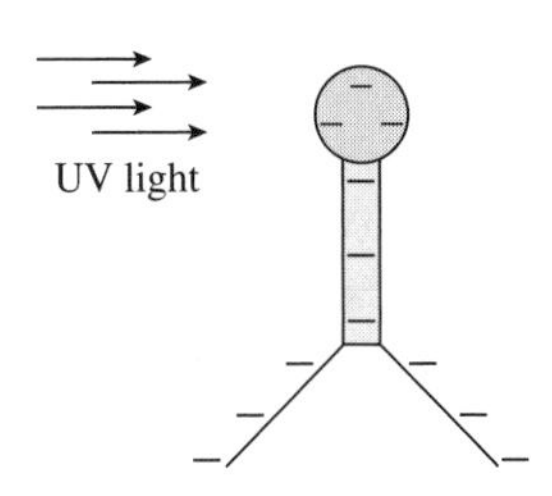

b. You might think that an ultraviolet light shining on an initially uncharged electroscope would cause the electroscope to become positively charged as photoelectrons are emitted. In fact, ultraviolet light has no noticeable effect on an uncharged electroscope. Why not?

3. Draw the trajectories of several typical photoelectrons when (a) $\Delta V = V_{\text{anode}} - V_{\text{cathode}} > 0$, (b) $-V_{\text{stop}} < \Delta V < 0$, and (c) $\Delta V < -V_{\text{stop}}$.

a.

Light

Cathode

Anode

A

$-$ $+$

$\Delta V > 0$

b.

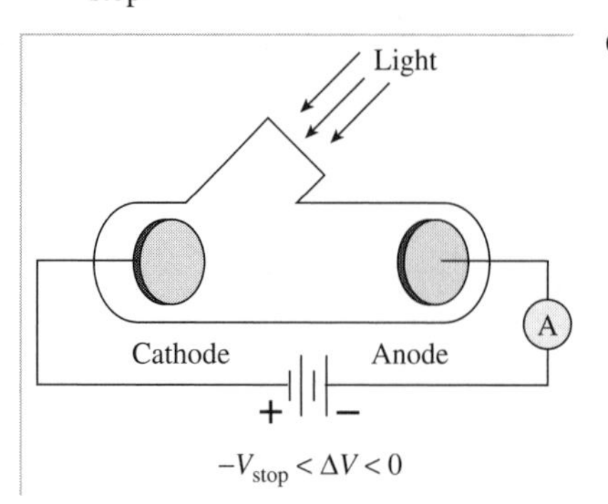

c.

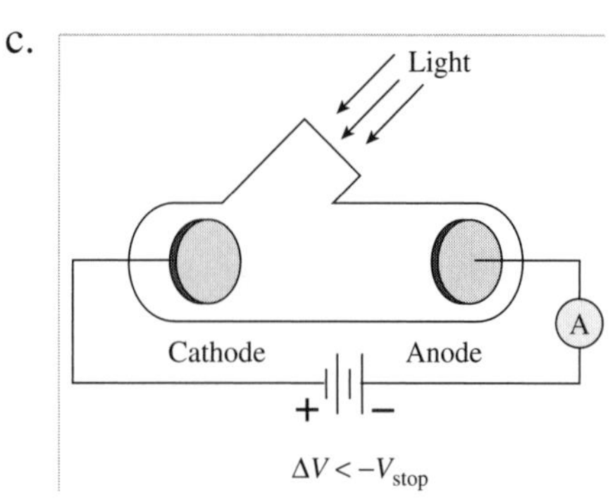

4. The work function of a metal measures:
   i. The kinetic energy of the electrons in the metal.
   ii. How tightly the electrons are bound within the metal.
   iii. The amount of work done by the metal when it expands.

   Which of these (perhaps more than one) are correct? Explain.

5. a. What is the significance of V_{stop}? That is, what have you learned if you measure V_{stop}? **Note:** Don't say that "$-V_{\text{stop}}$ is the potential that causes the current to stop." That is merely the definition of V_{stop}. It doesn't say what the *significance* of V_{stop} is.

   b. Why is it surprising that V_{stop} is independent of the light intensity? What would you *expect* V_{stop} to do as the intensity increases? Explain.

   c. If the wavelength of the light in a photoelectric effect experiment is increased, does V_{stop} increase, decrease, or stay the same? Explain.

6. The figure shows a typical current-versus-potential difference graph for a photoelectric effect experiment. On the figure, draw and label graphs for the following three situations:
   i. The light intensity is increased.
   ii. The light frequency is increased.
   iii. The cathode work function is increased.

   In each case, no other parameters of the experiment are changed.

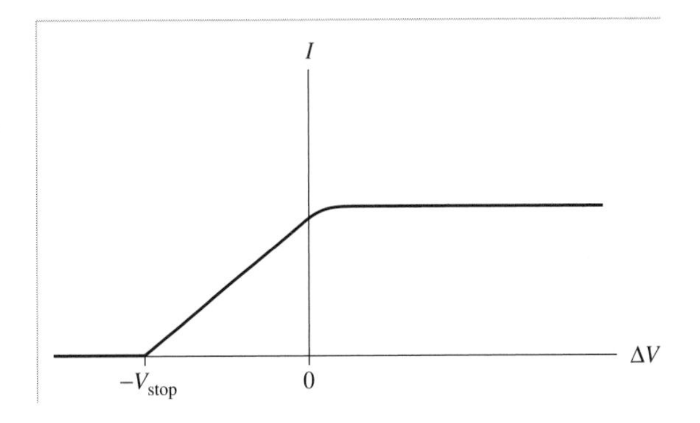

7. The figure shows a typical current-versus-frequency graph for a photoelectric effect experiment. On the figure, draw and label graphs for the following three situations:
   i. The light intensity is increased.
   ii. The anode-cathode potential difference is increased.
   iii. The cathode work function is increased.

   In each case, no other parameters of the experiment are changed.

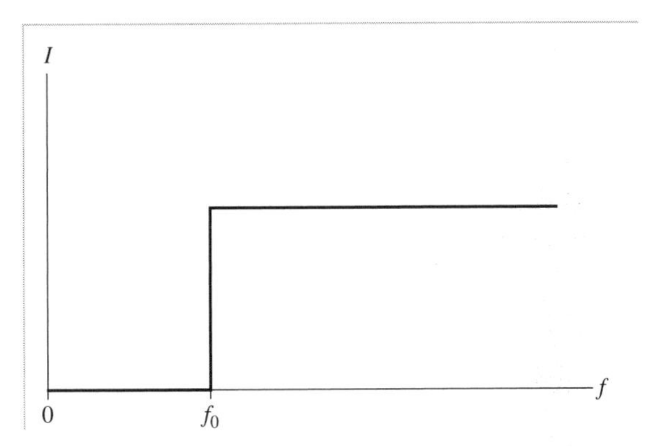

8. The figure shows a typical stopping potential-versus-frequency graph for a photoelectric effect experiment. On the figure, draw and label graphs for the following two situations:
   i. The light intensity is increased.
   ii. The cathode work function is increased.

   In each case, no other parameters of the experiment are changed.

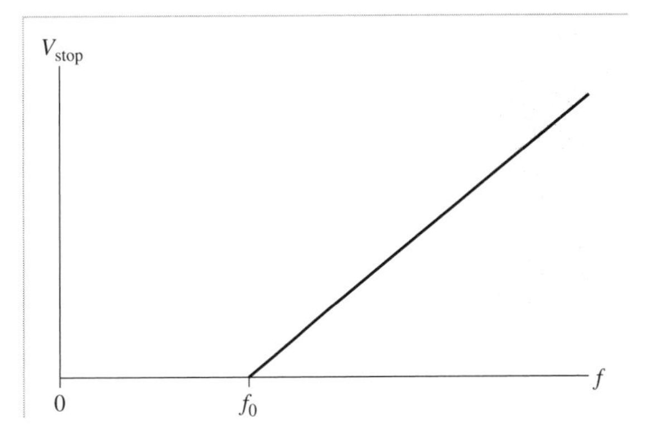

28.3 Photons

9. The top figure is the *negative* of the photograph of a single-slit diffraction pattern. That is, the darkest areas in the figure were the brightest areas on the screen. This photo was made with an extremely large number of photons.

   Suppose the slit is illuminated by an extremely weak light source, so weak that only 1 photon passes through the slit every second. Data are collected for 60 seconds. Draw 60 dots on the empty screen to show how you think the screen will look after 60 photons have been detected.

10. Photons are sometimes represented by pictures like this. Just what is this a picture of? Explain what this "graph" shows.

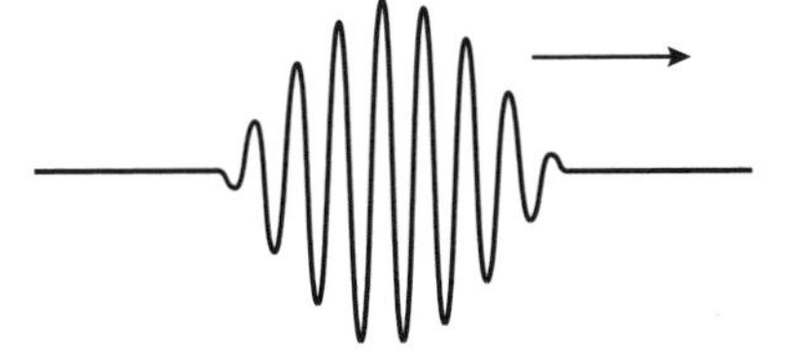

11. Light of wavelength $\lambda = 1\ \mu m$ is emitted from point A. A photon is detected $5\ \mu m$ away at point B. On the figure, draw the trajectory that a photon follows between points A and B.

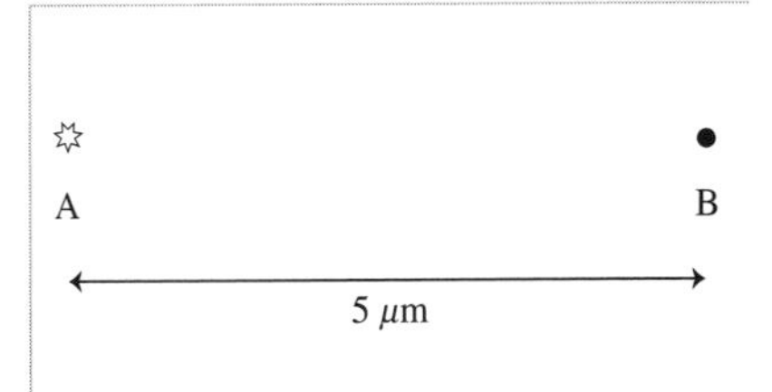

28.4 Matter Waves

12. The figure is a simulation of the electrons detected behind a very narrow double slit. Each bright dot represents one electron. How will this pattern change if the following experimental conditions are changed? Possible changes you should consider include the number of dots and the spacing, width, and positions of the fringes.

    a. The electron-beam intensity is increased.

    b. The electron speed is reduced.

    c. The electrons are replaced by positrons with the same speed. Positrons are antimatter particles that are identical to electrons except that they have a positive charge.

    d. One slit is closed.

13. Very slow neutrons pass through a single, very narrow slit. Use 50 or 60 dots to show how the neutron intensity will appear on a neutron-detector screen behind the slit.

14. Electron 1 is accelerated from rest through a potential difference of 100 V. Electron 2 is accelerated from rest through a potential difference of 200 V. Afterward, which electron has the larger de Broglie wavelength? Explain.

28.5 Energy Is Quantized

15. a. For the first few allowed energies of a particle in a box to be large, should the box be very big or very small? Explain.

b. Which is likely to have larger values for the first few allowed energies: an atom in a molecule, an electron in an atom, or a proton in a nucleus? Explain.

28.6 Energy Levels and Quantum Jumps

16. An electron in a quantum system is seen to emit photons of the following wavelengths: 248 nm, 310 nm, 496 nm, 827 nm, and 1240 nm. Draw an energy level diagram and indicate the transitions associated with each of these photon emissions if the lowest energy state for the electron is 2.0 eV.

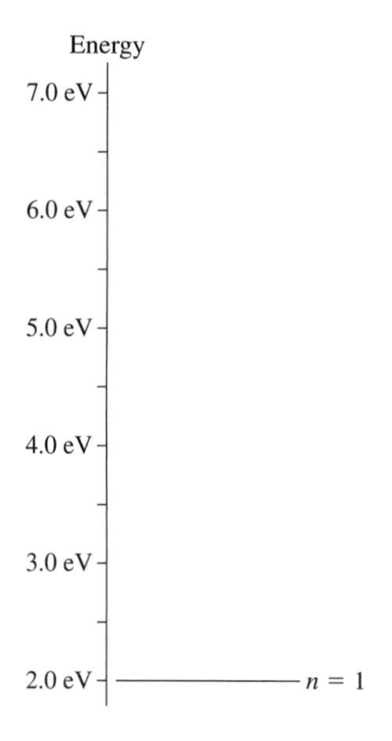

28.7 The Uncertainty Principle

17. Electrons of an initial momentum p_y diffract when passing through a single slit to form a central diffraction maximum along the x-direction on an array of detectors behind the slit. If electrons having twice the initial momentum were directed toward the same slit, would the resulting pattern be narrower, wider, or the same width? Explain.

28.8 Applications and Implications of Quantum Theory

No exercises for this section.

29 Atoms and Molecules

29.1 Spectroscopy

1. The figure shows the spectrum of a gas discharge tube.

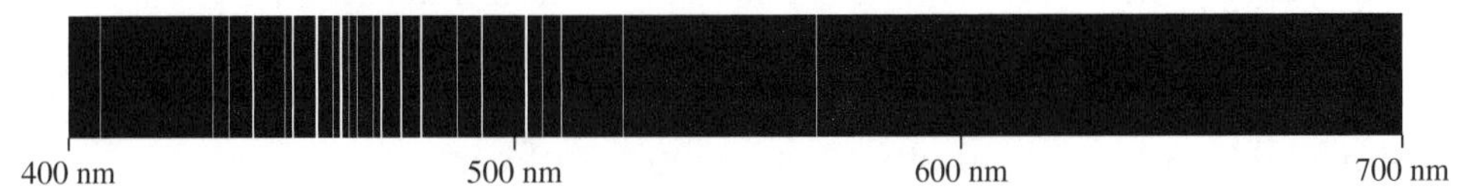

What color would the discharge appear to your eye? Explain.

2. A photograph of an absorption spectrum appears white with black lines. An emission spectrum appears black with bright lines. Why are they different?

29.2 Atoms

3. Suppose you throw a small, hard rubber ball through a tree. The tree has many outer leaves, so you cannot see clearly into the tree. Most of the time your ball passes through the tree and comes out the other side with little or no deflection. On occasion, the ball emerges at a very large angle to your direction of throw. On rare occasions, it even comes straight back toward you. From these observations, what can you conclude about the structure of the tree? Be specific as to how you arrive at these conclusions *from the observations*.

4. Beryllium is the fourth element in the periodic table. Draw pictures similar to Figure 29.9 showing the structure of neutral Be, of Be^+, of Be^{++}, and of the negative ion Be^-.

Be | Be^+ | Be^{++} | Be^-

5. The element hydrogen has three isotopes. The most common has $A = 1$. A rare form of hydrogen (called *deuterium*) has $A = 2$. An unstable, radioactive form of hydrogen (called *tritium)* has $A = 3$. Draw pictures similar to Figure 29.11 showing the structure of these three isotopes. Show all the electrons, protons, and neutrons of each.

$A = 1$ | $A = 2$ | $A = 3$

6. Tritium, the $A = 3$ isotope of hydrogen, is radioactive. One of the neutrons undergoes the transformation $n \rightarrow p^+ + e^- + \nu$, where ν is a massless, chargeless subatomic particle called a *neutrino*. Both the electron and the neutrino are ejected from the nucleus at high speed, but the proton remains. This is called *beta decay* of a nucleus.

   a. Draw a picture of the structure of a tritium atom immediately before and immediately after it undergoes beta decay. Show all the electrons, protons, and neutrons.

   b. Identify the element, the isotope (the A-value), and the charge state (neutral or singly charged) of the atom after the decay. Give your answer in symbolic form, such as $^6Li^+$.

7. Identify the element, the isotope, and the charge state. Give your answer in symbolic form, such as $^4He^+$ or $^8Be^-$.

a.

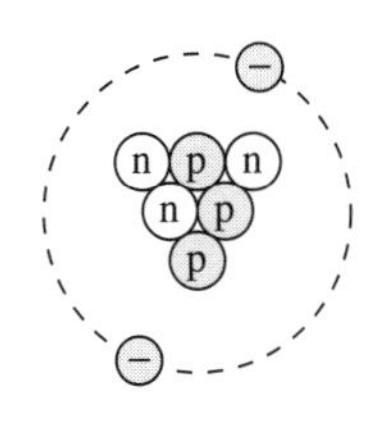

b.

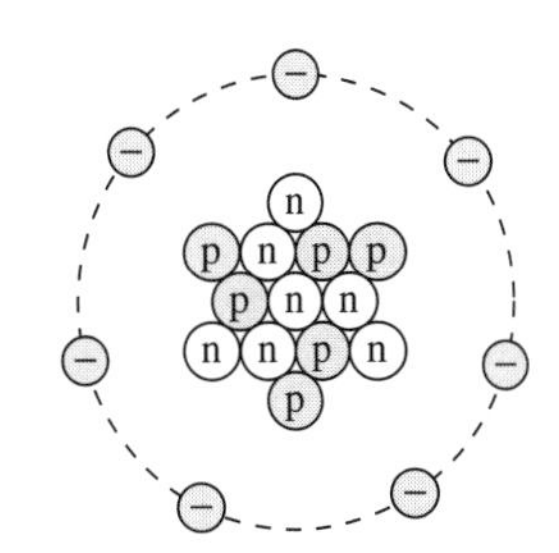

29.3 Bohr's Model of Atomic Quantization

29.4 The Bohr Hydrogen Atom

8. The figure shows a hydrogen atom, with an electron orbiting a proton.

   a. What force or forces act on the electron?

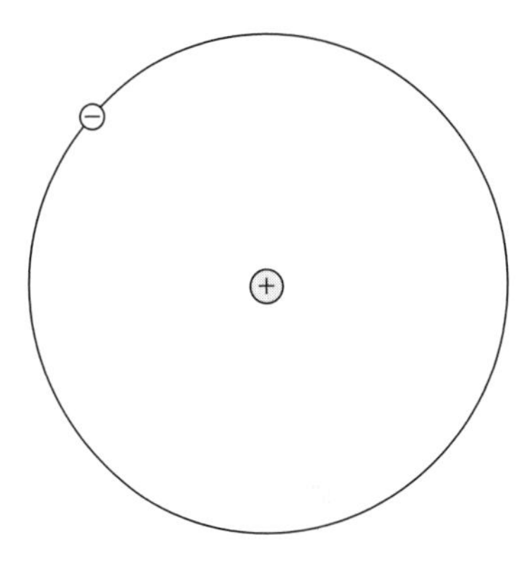

   b. On the figure, draw and label the electron's velocity, acceleration, and the force vectors.

9. a. The stationary state of hydrogen shown on the left has quantum number $n =$ ______

   b. On the right, draw the stationary state of the $n - 1$ state.

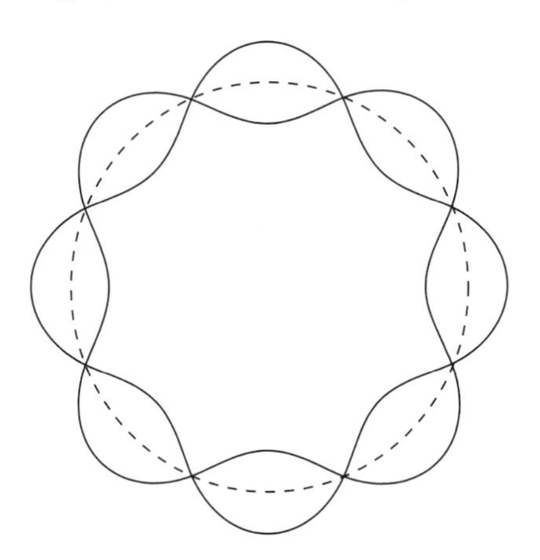

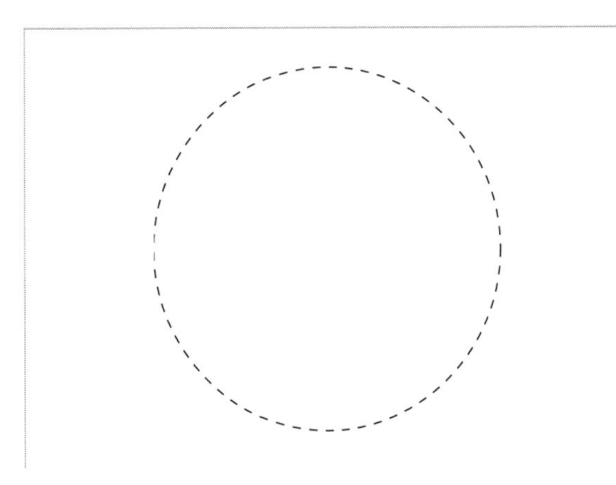

10. Why is there no stationary state of hydrogen with $E = -9$ eV?

11. Draw and label an energy level diagram for hydrogen. On it, show all the transitions by which an electron in the $n = 4$ state could emit a photon.

12. The longest wavelength in the Balmer series is 656 nm.

a. What transition is this?

b. If light of this wavelength shines on a container of hydrogen atoms, will the light be absorbed? Why or why not?

29.5 The Quantum-Mechanical Hydrogen Atom

13. a. How many electrons, protons, and neutrons are in a hydrogen-like ^{12}C ion?

b. Draw a picture of a hydrogen-like ^{12}C ion, showing all the particles you identified in part a.

14. a. As a multiple of $\hbar$, what is the angular momentum of a *d* electron? ______

b. What is the *maximum* *z*-component of angular momentum of a *d* electron? ______

c. Is $(L_z)_{max}$ greater than, less than, or equal to L? ______

d. What is the significance of your answer to part c?

29.6 Multielectron Atoms

15. Consider a 2*s* electron, as portrayed in Figures 29.21 and 29.25. In your own words, describe how these figures suggest a "shell structure" of electrons around the nucleus.

29.7 Excited States and Spectra

16. The figure shows the energy levels of a hypothetical atom.

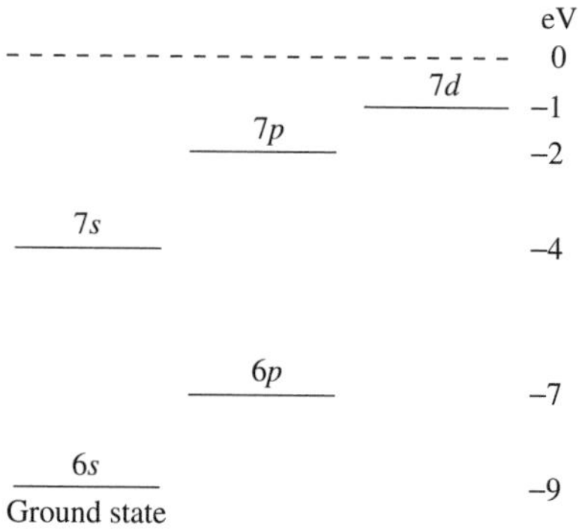

    a. What is the atom's ionization energy?

    b. In the space below, draw the energy-level diagram as it would appear if the ground state were chosen as the zero of energy. Label each level and the ionization limit with the appropriate energy.

29.8 Molecules

17. Laundry detergent manufacturers sometimes use fluorescent chemicals to make clothes appear whiter in the presence of small amounts of UV light. What role does UV light (which is not directly visible) play in making these "whites appear whiter"?

29.9 Stimulated Emission and Lasers

No exercises for this section.

30 Nuclear Physics

30.1 Nuclear Structure

1. Consider the atoms ^{16}O, ^{18}O, ^{18}F, ^{18}Ne, and ^{20}Ne. Some of the questions about these atoms may have more than one answer. Give all answers that apply.

   a. Which atoms are isotopes? __________

   b. Which atoms have the same number of nucleons? __________

   c. Which atoms have the same chemical properties? __________

   d. Which atoms have the same number of neutrons? __________

   e. Which atoms have the same number of valence electrons? __________

30.2 Nuclear Stability

2. a. Is the total binding energy of a nucleus with $A = 200$ more than, less than, or equal to the binding energy of a nucleus with $A = 60$? Explain.

   b. Is a nucleus with $A = 200$ more tightly bound, less tightly bound, or bound equally tightly as a nucleus with $A = 60$? Explain.

3. Binding energy calculations usually involve comparing two almost equal numbers, which requires careful attention to significant figures. In Chapter 1, you learned that numbers starting with a leading 1 should be given with an additional digit compared to other numbers. Compare the number of significant figures required to estimate the mass difference between a neutron and a proton to two significant figures if

a. the masses are given in atomic mass units (u).

b. the masses are given in units of MeV/c^2.

30.3 Forces and Energy in the Nucleus

4. Draw energy-level diagrams showing the nucleons in ^{6}Li and ^{7}Li.

a.

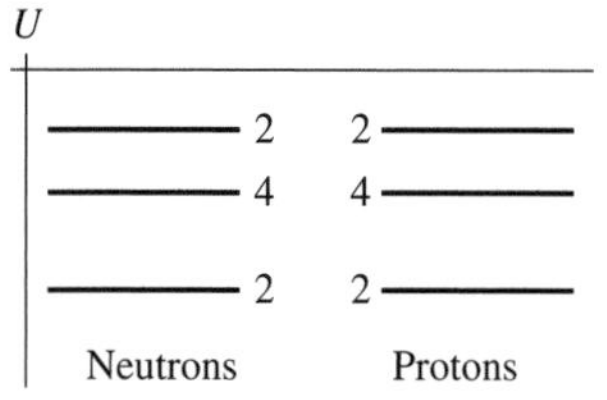

U

Neutrons		Protons
2		2
4		4
2		2

^{6}Li

b.

U

Neutrons		Protons
2		2
4		4
2		2

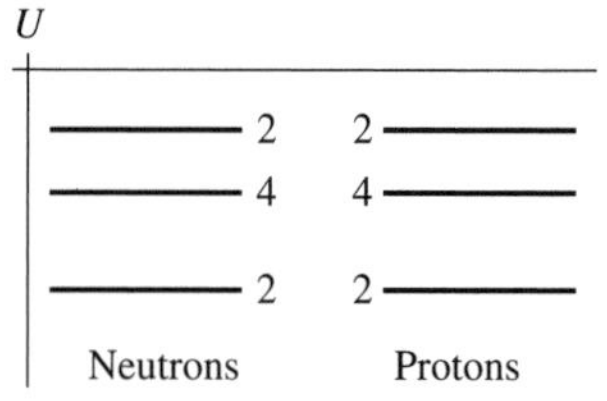

^{7}Li

30.4 Radiation and Radioactivity

5. The graph shows the activity (R) versus the number of nuclei (N) for a radioactive isotope A. Draw and label as B a line to represent the activity of an isotope with a half-life that is twice as long as that of A. Draw and label as C a line to represent the activity of an isotope with a half-life that is one-third as long as that of A.

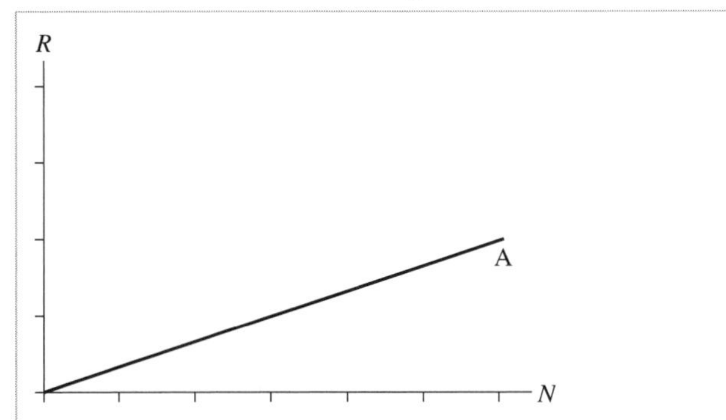

6. The half-life of a radioactive nucleus is 6 hours.

   a. Plot the number of nuclei remaining as a function of time on the axes below, assuming there were 1,000,000 nuclei at time $t = 0$.

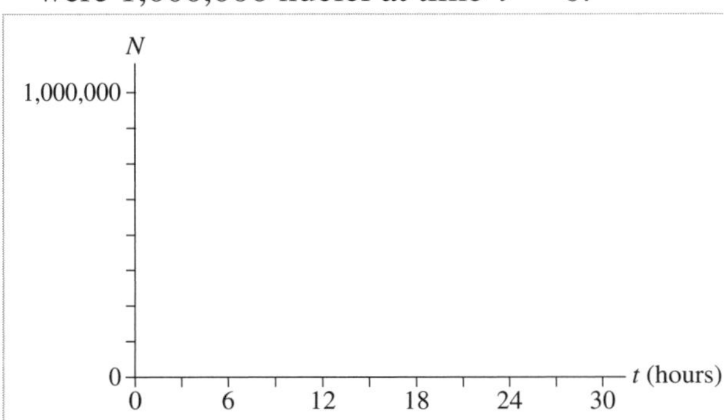

   b. Indicate the time constant τ for the decay by labeling it on the graph above.

   c. How many nuclei remain after one day? ______________

d. If the decay-product nuclei of the radioactive decay described in part a are stable, plot the number of decay-product nuclei as a function of time on the axes below.

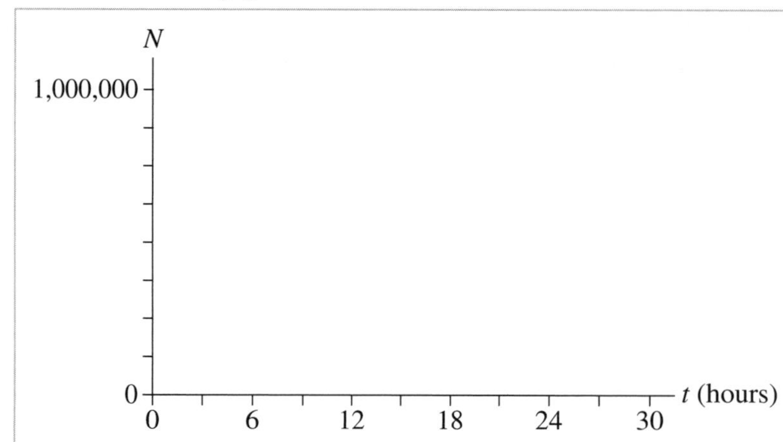

7. Use the properties of exponents to explain why it is difficult to use radioactive dating if the particular isotope's half-life ($t_{1/2}$) is, for example, two orders of magnitude shorter than the age (t) of the sample.

30.5 Nuclear Decay Mechanisms

8. An alpha particle is ejected from a nucleus with 6 MeV of kinetic energy. Was the alpha particle's kinetic energy inside the nucleus more than, less than, or equal to 6 MeV? Explain.

9. Part of the ^{236}U decay series is shown.
   a. Complete the labeling of atomic numbers and elements on the bottom edge.
   b. Label each arrow to show the type of decay.

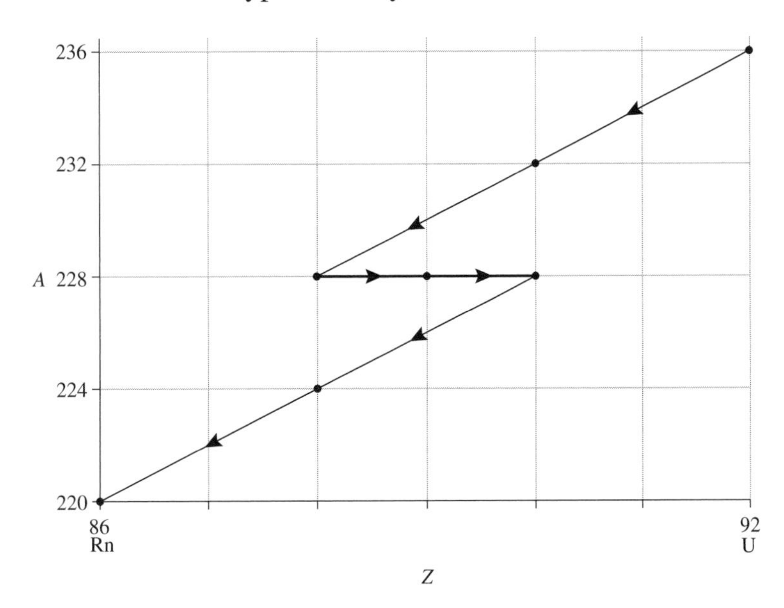

30.6 Subatomic Particles

30.7 Medical Applications of Nuclear Physics

10. What is the difference, or is there a difference, between radiation *dose* and *dose equivalent?*

Notes

Notes

Notes

Notes

Notes

Notes